Lecture Notes in Mathematics

Edited by A. Dold and B. Eckmann

852

Laurent Schwartz

Geometry and Probability in Banach Spaces

Notes by Paul R. Chernoff

Springer-Verlag
Berlin Heidelberg New York 1981

Author

Laurent Schwartz
Centre de Mathématiques de l'Ecole Polytechnique
91128 Palaiseau Cedex, France

Paul R. Chernoff
Dept. of Mathematics, University of California
Berkeley, CA 94720, USA

AMS Subject Classifications (1980): 46 B 20, 47 B 10, 60 B 11

ISBN 3-540-10691-X Springer-Verlag Berlin Heidelberg New York
ISBN 0-387-10691-X Springer-Verlag New York Heidelberg Berlin

Printing and binding: Beltz Offsetdruck, Hemsbach/Bergstr.
2141/3140-543210

PREFACE

These Notes correspond to a course of lectures I gave at the University of California, Berkeley, in April-May 1978. I tried to present, in these lectures, the main results of geometry and probability in Banach spaces, which have been the material of several years of the Séminaire de l'Ecole Polytechnique. Difficult task! A lot of material in a short time! It was possible to state a great number of theorems, and to prove a large part of them. Of course, the longest proofs have been omitted. However, I believe that somebody who seriously attended the lectures or who reads these Notes will be able to work by himself in this theory. I want to say that I was delighted by the atmosphere in the audience; people seemed to enjoy the lectures very much, and surely I enjoyed myself! Paul CHERNOFF gives here a very good account of the series of lectures, with a nice expression of his personal taste; I want to thank him very much!

INTRODUCTION

As said in the Preface, these Lectures summarize a great number of results given in several years of séminaires of the Ecole Polytechnique, Palaiseau, France. They cover relationships between geometrical properties, properties of functional analysis, probabilistic properties in Banach spaces, frequently appearing a priori as completely independent of each other. After the brilliant past of the Banach spaces with the Polish school, specially Banach, these spaces were a little abandonned for general locally convex topological vector spaces after World War II (in particular with applications to distributions, and Grothendieck's results about nuclear spaces). But Banach spaces came again, for more refined results, in the 60's, with Lindenstrauss, Pełczynski and others, turning around the L^p spaces ; the present subject goes into this direction. Many of the results given here have been found by mathematicians of the French school, in particular Bernard Maurey and Gilles Pisier.

Chapter I. Lecture 1 gives a rapid statement of the main ideas of the book. One starts with Lecture 2 with p-summing maps, studied first by Pietsch : a map is p-summing, $-1 < p \leq +\infty$, if it transforms every scalarly ℓ^p-sequence into a true ℓ^p-sequence : it may be characterized by Pietsch inequality, or Pietsch factorization (fundamental role of the natural injection $L^\infty \to L^p$ with respect to a probability measure). Lecture 3 gives applications. If a map is p-summing, it is q-summing for $q \geq p$ (going up phenomenon) ; therefore there is a cut $p_o(u)$ such that the map u is p-summing for every $p > p_o$, and is not for $p < p_o$. If $p(u) = +\infty$, u is just continuous, nothing more ; if $p_o(u) = -1$, u is p-summing for every p, or completely summing. It is announced (it will be proved only later on) that a map between Hilbert spaces is p-summing for p finite iff it is Hilbert-Schmidt (Pełczynski) ; here $p_o(u)$ is always -1 or $+\infty$. But there is also a very interesting going down phenomenon, Pietsch' conjecture (proved in 1972 by Simone Chevet and Maurey) : if u is $(1-\varepsilon)$-summing, $\varepsilon > 0$, it is completely summing. One says that E is p-Pietsch if every map from E into every Banach space, which is p-summing, is completely summing ; every Banach space is $(1-\varepsilon)$-Pietsch. The Sup of the p such that E is p-Pietsch is Pietsch (E), it it will be studied later on. Lecture 4 studies p-integral and p-nuclear maps, Hilbert-Schmidt maps for Hilbert spaces.

Chapter II. Cylindrical probabilities and radonifying maps.
Lecture 4 (continued) : definition of Radon measures on topological spaces, of cylindrical probabilities on locally convex topological vector spaces (coherent

systems of probabilities on the finite dimensional factor spaces), gaussian measures on Hilbert spaces, Prokhorov condition for a cylindrical probability to be Radon.

Lecture 5. It gives the duality theorem for Hilbert spaces, giving necessary and sufficient conditions for a map between Hilbert spaces to be p-summing (announced in Lecture 3). One defines p-radonifying maps, narrow topology for Radon measures and cylindrical topology for cylindrical probabilities.

Lecture 6 gives relationships between p-summing and p-radonifying maps : u is p-summing iff it is approximatively p-radonifying into the bidual. Special good case ; $1 < p < +\infty$, where there is equivalence between p-summing and p-radonifying. Lecture 7 introduces P. Lévy's p-stable, or p-gauss laws, $0 < p \leq 2$. Then Lecture 8 and theorem 8.2 give the general duality theorem for p-gauss laws, generalizing the results of lecture 5. It extends, for $p < 1$, to the proof of Pietsch' conjecture. Lecture 9 introduces p-Pietsch spaces ; and studying radonification between L^p spaces and Sobolev spaces, it proves, by the theory of p-radonifying maps, the continuity and Hölder properties of the Brownian motion. Lecture 10 ties cylindrical probabilities and linear stochastic processes, proving that L^q can be embedded into L^p for $0 \leq p \leq q \leq 2$.

Chapter III. Type and cotype.

Here we introduce probabilistic properties of Banach spaces. It has been known for a long time that if $(\varepsilon_n)_{n\in\mathbb{N}}$ are independent random variables, with values ± 1 with probability 1/2, and if the x_n are given complex numbers, the series $\sum_n \varepsilon_n x_n$ is almost surely convergent if $(\sum |x_n^2|)^{1/2} < +\infty$, and almost surely divergent in the opposite case. What about the case of given x_n in a Banach space E ? The same result is never true, except if E can be renormed as to become Hilbert. E is said to be of type p, $1 \leq p \leq 2$, if $(\sum_n |x_n|^p)^{1/p} < +\infty$ implies that $\sum_n \varepsilon_n x_n$ is almost surely convergent, of cotype q, $+\infty \geq q \geq 2$, if the almost sure convergence of $\sum \varepsilon_n x_n$ implies $(\sum_n |x_n|^q)^{1/q} < +\infty$. Lecture 11 gives Kahane inequality for the sums $\sum_n \varepsilon_n x_n$, and proves that, for $1 \leq r \leq 2$, L^r has type r and cotype 2, and nothing better, and, for $2 \leq r < +\infty$, type 2 and cotype r, and nothing better, while L^∞ is very bad, type 1 and cotype $+\infty$. Lecture 12 gives Pisier's relationships between types defined by the Rademacher variables ε_n and other random variables Z_n ; it defines p-gauss type, very important. It gives relationships between Rademacher type and Gauss type : E is of type p-gauss iff it is of type $(p+\varepsilon)$-Rademacher, for $p < 2$, while, for $p = 2$, both types are equivalent. One calls Type (E) the supremum of the p's such that E has type p, and Cotype (E) the infimum of the q's such that E has cotype q ; Type (E) $\leq 2 \leq$ Cotype (E).

Chapter IV. Ultrapowers and superproperties.

(This long chapter is the last one.) Lecture 12 (continued) defines the finite representativity : F is finitely representable into E if, for every finite dimensional subspace F_1 of F, and every $\varepsilon > 0$, there exists a finite dimensional subspace E_1 of E which is $(1+\varepsilon)$-isomorphic to F_1 ; E contains almost copies of all the finite-dimensional subspaces of F. If P is a property of Banach spaces, Super P is the following one : E has Super P if every Banach F, finitely representable into E, has P (hence also Super P). P is a superproperty if Super P = P. Dvoretzky's theorem says that L^2 is finitely representable into every infinite dimensional Banach space ; S(E) will denote the set of p's for which L^p is finitely representable into E, hence $2 \in S(E)$. Lecture 12 states also Maurey-Pisier's result : E has type gauss $p < 2$, iff $p \notin S(E)$. Lecture 13 studies the Maurey factorization theorems, in relationship with type and cotype. It finds again the famous Grothendieck factorization property of a map $L^q \to L^p$ through L^2, for $p \leq 2 \leq q$. Lecture 14 states relationships between factorization properties and p-Pietsch properties. Lecture 15 introduces ultrapowers $E^I/\mathcal{U}$ of a Banach space E, with respect to an ultrafilter $\mathcal{U}$ on the set of indices I ; F is finitely representable into E iff it is a subspace of an ultrapower of E. Lecture 16 gives the fundamental relationships between S(E) and Pietsch (E). Unifying all the previous results in this direction, one gets the final Maurey-Pisier-Krivine theorem, which leads back 5 numbers to be 2 only : Min S(E) = Type (E), Max S(E) = Cotype (E) = Hölder conjugate of Pietsch (E) (•). Pisier improved even this result in July 1980 : except in the case Type (E) = Type (E') = 1, one has always : Cotype (E') is the Hölder conjugate of Type (E) (see page 97). Lecture 16 also studies various equivalent forms of the weaker non trivial superproperty of E : $+\infty \notin S(E)$. Lecture 17 studies Gauss-summing maps, the special properties of the Banach spaces of cotype 2, and a refinement of an old Grothendieck deep result : every map of L^1 into a Hilbert space is completely summing. Lectures 18 and 19 study superreflexive Banach spaces, and give (theorem 19.5) 12 equivalent properties. Theorem 19.6 gives, in a very simple way by the previous results, the properties of the subspaces of L^p, $1 \leq p < 2$, given by Rosenthal.

Laurent SCHWARTZ, December 1980

(•) We don't give any bibliography in this text ; it would be too abundant, and the "Séminaires de l'Ecole Polytechnique" contain a lot of references. However let us point out that this main result of lecture 16 is proved in an almost self-contained article of Bernard Maurey and Gilles Pisier : "Séries de variables aléatoires vectorielles indépendantes, et propriétés géométriques des espaces de Banach", Studia Math. 58 (1976), 45-90.

CONTENTS

I. INTRODUCTION

Lecture 1. Type and Cotype for a Banach Space p-Summing Maps

Let E be a Banach space. We may distinguish *functional-analytic, geometric,* and *probabilistic* properties of E. And we shall study their relationships.

Examples: *Reflexivity* of E is a functional-analytic property.

Uniform convexity is a geometric property. Recall that E is uniformly convex provided that $(\forall\varepsilon>0)(\exists\delta>0)$ so that whenever $|x| = |y| = 1$ and $|x-y| \geq \varepsilon$ it follows that $|\frac{x+y}{2}| \leq 1 - \delta$.

A well known Theorem: A uniformly convex Banach space is reflexive. (The converse is false.)

We turn now to probabilistic properties, related to the notions of "type" and "cotype" which will be central importance for us.

Let $(x_n)_{n\in\mathbb{N}}$ be a sequence of *real numbers*. Let ε_n be a sequence of "random signs", i.e. independent identically distributed random variables (R.V.'s) taking the values $+1$, -1 with probability 1/2. Consider the sum $\sum_{n=0}^{\infty} \varepsilon_n x_n$. Then, according to a well known theorem in probability theory, the following are equivalent:

(i) $\sum_n \varepsilon_n x_n$ converges almost surely (a.s.)

(ii) $\sum_n |x_n|^2 < +\infty$.

We may now consider the same situation with the real numbers being replaced by a Banach space E: thus $(x_n)_{n\in\mathbb{N}}$ is a sequence of vectors in E, while the R.V.'s ε_n are as above. (These R.V.'s are often called "Rademacher variables".) We ask: are (i) and (ii) still equivalent" (Of course $|x|$ denotes the norm of a vector x.) In fact properties (i) and (ii) are equivalent if and only if E is "Hilbertizable" (isomorphic to a Hilbert space).

Which Banach spaces have the property that (ii) $\Rightarrow$ (i)? We call them *type 2* spaces. More generally, we say that E is of *type p* (for $0 < p \leq 2$) provided that in E

$$(*) \qquad \sum_{n=0}^{\infty} |x_n|^p < +\infty \Rightarrow \sum_{n=0}^{\infty} \varepsilon_n x_n \text{ converges a.s.}$$

Note: It is pointless to consider $p > 2$, for only $E = (0)$ is then of type p. For if the dimension is positive we can take all the x_n to lie in some one-dimensional subspace, and then $\mathbb{R}$ would be of type p, $p > 2$, which is false (because an ℓ^p sequence need not be ℓ^2).

On the other hand, it is completely trivial that every Banach space is *type* 1 (since an absolutely convergent series certainly converges). So we need only consider the range $1 \leq p \leq 2$.

Example: Consider *infinite dimensional* L^r spaces. Then

for $1 \leq r \leq 2$, L^r is type r (and no better);

for $2 \leq r < +\infty$, L^r is type 2;

for $r = +\infty$, L^∞ is type 1.

We shall see that "type" is *inherited* by subspaces and quotient spaces. This again shows that L^1 and L^∞ must be type 1 (and not better than type 1) since every Banach space is a subspace of an L^∞ and a quotient of an L^1.

The notion of cotype arises by "reversing" (*): we say that E is of *cotype q*, $2 \leq q \leq +\infty$, provided that

$$(**) \qquad \sum_{n=0}^{\infty} \varepsilon_n x_n \text{ converges a.s.} \Rightarrow \sum_{n=0}^{\infty} |x_n|^q < +\infty$$

(with the obvious modification if $q = \infty$, namely $\sup_n |x_n| < +\infty$). Every Banach space is of cotype $+\infty$, and no space except (0) can have cotype < 2. Cotype is inherited by subspaces, but *not* necessarily by quotient spaces. So L^∞ (if infinite dimensional) cannot be of cotype better than $+\infty$. But L^1 is of cotype 2. This has important applications.

There are relationships among these properties and the other analytic and geometric properties of Banach spaces. For example, let $\delta(\varepsilon)$ be the modulus of (uniform) convexity. Then if $\delta(\varepsilon) \sim \varepsilon^q$, $q \geq 2$, the space has cotype q.

We turn now to the subject of *p-summing maps* (developed by Pietsch and Kwapien).

Consider a sequence $e = (e_n)_{n\in\mathbb{N}}$ of vectors in a Banach space E. We say that e is in $\ell^p(E)$ provided $\sum_n |e_n|^p < +\infty$, and we write $\|e\|_p = (\sum_n |e_n|^p)^{1/p}$ (with the usual modification if $p = +\infty$). Note that we are considering the range $0 < p \leq +\infty$. The space $\ell^p(E)$ is a complete metric space, though not a normed space if $p < 1$. Obviously we have the inclusion $\ell^q(E) \subseteq \ell^p(E)$ if $q \leq p$.

Definition: The sequence e is *scalarly* ℓ^p, and we write $e \in S\ell^p(E)$, provided that for every $\xi \in E'$ the sequence $(\langle \xi, e_n\rangle)_{n\ \mathbb{N}}$ belongs to ℓ^p, i.e. $\sum_n |\langle \xi, e_n\rangle|^p < +\infty$.

Obviously $e \in \ell^p(E) \Rightarrow e \in S\ell^p(E)$, but the converse is false. For consider (e_n) an orthonormal basis of Hilbert space H. Then $e = (e_n) \in \ell^\infty(H)$ only, but e is *scalarly* ℓ^2.

Of course, if E is finite dimensional, then ℓ^p and scalarly ℓ^p are the same. But (except for $p = \infty$) in *every* infinite-dimensional Banach space, and for every $p < \infty$, there exist scalarly ℓ^p sequences which are not ℓ^p. When $p = \infty$, the Banach-Steinhaus Theorem tells us that scalarly $\ell^\infty \Rightarrow \ell^\infty$.

More generally, we have that a necessary and sufficient condition for e to be $S\ell^p$ is:

$$\sup_{|\xi|\leq 1} \sum_{n=0}^{\infty} |\langle \xi, e_n\rangle|^p < +\infty.$$

Proof: We have already dealt with the case $p = \infty$. The condition is trivially sufficient. Conversely, $e \in S\ell^p(E)$ defines a linear map from E' to ℓ^p via $\xi \mapsto (\langle \xi, e_n\rangle)_{n\in\mathbb{N}}$. This map is continuous because it is the pointwise limit of its finite-rank sections (here we use $p < +\infty$). Hence the above supremum is finite.

Having made this observation we can define the "scalarly ℓ^p norm" by

$$\|e\|_p^* = \sup_{|\xi|\leq 1} \left(\sum_{n=0}^{\infty} |\langle \xi, e_n\rangle|^p\right)^{1/p}.$$

Obviously $\|e\|_p^* \leq \|e\|_p$, while $\|e\|_\infty^* = \|e\|_\infty$.

Consider a continuous linear map $u: E \to F$. Obviously u maps ℓ^p sequences to ℓ^p sequences, with $\|u(e)\|_p \leq \|u\|\|e\|_p$. Also u takes $S\ell^p$ sequences to $S\ell^p$

sequences. Indeed,

$$\sum_n |\langle u(e_n),\eta\rangle|^p = \sum_n |\langle e_n, {}^t u(\eta)\rangle|^p$$

where ${}^t u$ is the transpose of u. Hence we get

$$\|u(e)\|_p^* \leq \|u\|\|e\|_p^*.$$

<u>Definition</u>: u is *p-summing* provided it carries scalarly ℓ^p sequences into ℓ^p sequences. (There is a related notion of "p,q-summing".) In this case there exists a constant $C < +\infty$ so that

$$(\dagger) \qquad \|u(e)\|_p \leq C\|e\|_p^*$$

Conversely, suppose that C exists such that $(\dagger)$ holds for all *finite* sequences. Then it must hold for infinite sequences as well, by passage to the limit. This gives us an equivalent definition of p-summing map.

The best constant in $(\dagger)$ is denoted $\pi_p(u)$ and is called the *p-summing norm* of u.

<u>Some facts</u> (to be proved below)

(1) If $q \geq p$ and u is p-summing, then u is q-summing, and $\pi_q(u) \leq \pi_p(u)$.

(2) Let (Z,ν) be a probability space. Then the canonical injection $L^\infty(Z,\nu) \to L^p(Z,\nu)$ is a p-summing map. In a sense this is the prototype of *all* p-summing maps: every p-summing map can be subfactored through this one in a sense we will see later on.

(3) Let E, F be Hilbert spaces, $u: E \to F$. Then u is 2-summing if and only if u is Hilbert-Schmidt. Moreover $\pi_2(u)$ equals the Hilbert-Schmidt norm of u.

A much deeper fact: if $p < +\infty$, then u is p-summing if and only if u is Hilbert-Schmidt. The proof uses a probabilistic argument involving Gauss laws. Moreover it turns out that $\pi_p(u)$ depends on the p-moments of the Gauss law. (So the use of Gauss laws here seems unavoidable.)

Question: in what Banach spaces are all the notions of p-summing equivalent?

(4) Let $u: E \to F$ be q-summing. If F is of cotype 2 and $q \geq 2$, then u is 2-summing. If E is of cotype 2, 2-summing implies p-summing for *all* p. So if both E and F are of cotype 2, all the notions of p-summing are equivalent. Example: $E = F = L^1$.

We will turn to the proofs of these facts in the next lecture.

Lecture 2. Pietsch Factorization Theorem

2.1. Theorem: Every linear map $u: E \to F$ is $+\infty$-summing. If u is p-summing and $q \geq p$ then u is q-summing, and moreover $\pi_q(u) \leq \pi_p(u)$. Hence there exists a p_0 (possibly $+\infty$) so that if $p > p_0$ then u is p-summing, while if $p < p_0$ then u is not p-summing; u may or may not be p_0-summing.

Proof: Since a sequence is scalarly bounded if and only if it is bounded, it is trivial that every u (assumed bounded) is $+\infty$-summing.

Suppose that u is p-summing and $q > p$. Let e be a scalarly ℓ^q sequence. Define r by the relation $\frac{1}{r} = \frac{1}{p} - \frac{1}{q}$. Then by Hölder's inequality, if $\alpha \in \ell^r$ then the sequence αe $(= \{\alpha_n e_n\})$ is scalarly ℓ^p; moreover, if $\|\alpha\|_r \leq 1$ then $\|\alpha e\|_p^* \leq \|e\|_q^*$. Therefore $\alpha u(e) = u(\alpha e)$ is a true ℓ^p sequence in F, with $\|\alpha u(e)\|_p \leq \pi_p(u)\|\alpha e\|_p^* \leq \pi_p(u)\|e\|_q^*$. But then we deduce from the converse of Hölder's inequality that $u(e)$ is an ℓ^q sequence with $\|u(e)\|_q = \sup\{\|\alpha u(e)\|_p : \|\alpha\|_r \leq 1\} \leq \pi_p(u)\|e\|_q^*$. Thus u is q-summing, and $\pi_q(u) \leq \pi_p(u)$. ∎

Note that for an "arbitrary" $u \in \mathcal{L}(E,F)$, p_0 will usually be $+\infty$. For example, if E is infinite-dimensional, the identity I is only $+\infty$-summing. Pietsch (around 1964) gave examples with $p_0 \geq 1$ arbitrary, and with u either p_0-summing or not. But Pietsch lacked examples with $p_0 < 1$. This led him to conjecture that either p_0 is ≥ 1 or else $p_0 = 0$. The Pietsch conjecture was proved (in 1972/73) by Simone Chevet and Bernard Maurey (independently).

Definition: u is *completely summing* if u is p-summing for all $p > 0$.

Thus the Pietsch conjecture says that if u is $1-\varepsilon$ summing, then u is completely summing. We will give the proof in Lecture 8.

The next Theorem provides an important insight into the structure of p-summing maps.

2.2. Pietsch Majorization Theorem: Let ν be a probability measure on a compact Hausdorff space Z. Suppose that $u\colon E \to F$ and $v\colon E \to C(Z)$ are bounded linear maps, and assume that there is a constant C such that we have the estimate, for all $x \in E$:

$$\begin{array}{ccc} E & \xrightarrow{v} & C(Z) \\ & \searrow^{u} & \\ & & F \end{array} \qquad \begin{aligned} \|u(x)\| &\leq C\|v(x)\|_{L^p(Z,\nu)} \\ &= C\left[\int_Z |v(x)(z)|^1 d\nu(z)\right]^{1/p} . \end{aligned}$$

Then u is p-summing, and $\pi_p(u) \leq C\|v\|$.

Proof: Consider any finite sequence $e_1, e_2, \ldots, e_n$ in E. Then

$$\begin{aligned} \sum_{i=1}^{n} \|u(e_i)\|^p &\leq C^p \sum_{i=1}^{n} \int_Z |v(e_i)(z)|^p \nu(dz) \\ &\leq C^p \sup_Z \sum_{i=1}^{n} |v(e_i)(z)|^p , \end{aligned}$$

where the last inequality uses the fact that ν is a probability measure. Now, for $z \in Z$, we have

$$v(e_i)(z) = \langle v(e_i), \delta_z \rangle = \langle e_i, {}^t v(\delta_z) \rangle ,$$

so that the right side of the inequality is

$$C^p \sup_Z \sum_{i=1}^{n} |\langle e_i, {}^t v(\delta_z) \rangle|^p .$$

Now ${}^t v(\delta_z)$ is a vector in E' with norm $\leq \|{}^t v\| \|\delta_z\| = \|{}^t v\| = \|v\|$. So the right side above is

$$\begin{aligned} &\leq C^p \|v\|^p \sup_{|\xi| \leq 1} \sum_{i=1}^{n} |\langle e_i, \xi \rangle|^p \\ &\leq C^p \|v\|^p \|e\|_p^* . \end{aligned}$$

Thus u is p-summing with $\pi_p(u) \leq C\|v\|$. ∎

Applications

(1) Suppose that we have a similar set-up but with $C(Z)$ replaced by $L^\infty(Z,\nu)$:

$$\begin{array}{ccc} E & \xrightarrow{v} & L^\infty(Z,\nu) \\ & \searrow^{u} & \\ & & F \end{array}$$

with $\|u(x)\| \leq C\|v(x)\|_{L^p}$ as above. Then we have the same conclusion: u is p-summing with $\|\pi_p(u)\| \leq C\|v\|$. This is immediate if we recall that $L^\infty(Z,\nu)$, being a commutative C*-algebra, is isomorphic to $C(\tilde{Z})$ where $\tilde{Z}$ is some "Stonian" compact Hausdorff space. Define $\tilde{\nu}$ on $\tilde{Z}$ to be the measure associated with the linear functional $\int_Z \phi\, d\nu$. Then $L^\infty(Z,\nu)$ is identified with $C(\tilde{Z})$ and $L^p(Z,\nu)$ with $L^p(\tilde{Z},\tilde{\nu})$.

The theorem says that the canonical embedding j of $L^\infty(Z,\nu)$ in $L^p(Z,\nu)$ is p-summing with π_p-norm at most 1. The result is true for $p < 1$ as well, although L^p is of course not Banach in that case. (This is a special case of the theorem, but it implies the general case.)

A natural question: when is the embedding of L^p in L^r $(p > r)$ q-summing? This is not as easy to answer.

(2) Observe that the p-summing maps form an ideal. That is, if *one* of u, v, w is p-summing then uvw is p-summing, and, e.g., $\pi_p(uvw) \leq \|u\|\pi_p(v)\|w\|$. So a map u with the following "factorization" is p-summing:

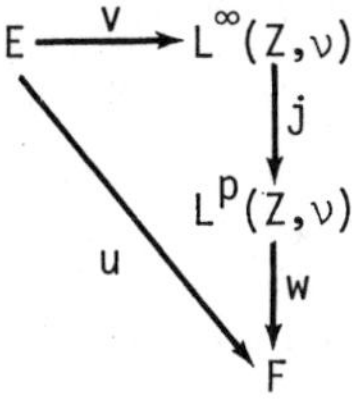

We can in fact draw a stronger conclusion from the theorem. The map w need not be defined on all of L^p; we only need to suppose it is defined on S, the closure of the range of $j \circ v$ in L^p.

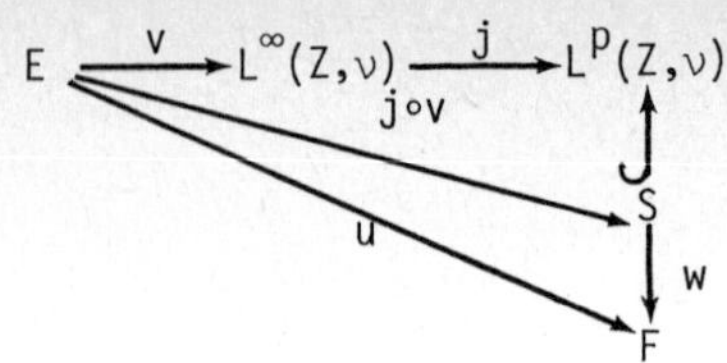

This we call a *sub-factorization* of u through L^p. Our theorem tells us that in this case u is p-summing. Such sub-factorizability is *exactly* equivalent to the existence of a Pietsch majorization inequality for u.

Indeed, the majorization $\|u(x)\| \leq C\|v(x)\|$ implies that $v(x) \mapsto u(x)$ defined as a map of norm $\leq C$ from $jv(E)$ into F, and it is therefore continuable as a map w from S into F, where S is the closure of $jv(E)$ in L^p. (Note that for $p = 2$ sub-factorizability implies true factorizability through L^2; but this is false for other values of p.)

The next Theorem establishes the converse of the majorization theorem, for $p < \infty$.

2.3. Theorem (Pietsch): Let p be finite. Suppose that $u: E \to F$ is p-summing. Then there exists a probability space (Z,ν) as above together with a continuous linear map $v: E \to C(Z)$ such that we have "majorization": $\|u(x)\| \leq C\|v(x)\|_{L^p}$. Moreover $\pi_p(u) = C\|v\|$ in this case.

Proof: Take Z to be B', the unit ball of E' with the weak* topology. Let $v: E \to C(B')$ be the canonical map, namely $v(x)(\xi) = \langle x,\xi\rangle$. Although v is independent of u, the probability measure ν must depend on u. We seek such a measure on B' so that, for all $x \in E$,

$$|u(x)| \leq \pi_p(u)[\int_{B'} |\langle x,\xi\rangle|^p \nu(d\xi)]^{1/p} \quad .$$

Our strategy will be to construct such a ν using the Hahn-Banach theorem.

Consider the collection Γ of all functions $\phi \in C(B')$ of the form

$$\phi(\xi) = \sum_{i=1}^{n} \|u(e_i)\|^p - \pi_p(u)^p \sum_{i=1}^{n} |\langle e_i,\xi\rangle|^p,$$

where $e_1,\ldots,e_n \in E$. Γ is closed under addition and admits multiplication by positive constants, so Γ is a convex cone in $C(B')$. Because u is p-summing with p-norm $\pi_p(u)$, the minimum of any $\phi \in \Gamma$ is *non-positive*.

Now let θ_+ be the open convex cone consisting of the strictly positive functions in $C(B')$. Note that $\theta_+ \cap \Gamma = \emptyset$. Hence the Hahn-Banach Theorem supplies us with a continuous linear functional which separates θ_+ and Γ, i.e. a non-zero Radon measure ν on B' such that $\int \phi d\nu > 0$ for $\phi \in \theta_+$ while $\int \phi d\nu \leq 0$ for $\phi \in \Gamma$. We may normalize ν to be a probability measure. Then we have $\int \phi\, d\nu \leq 0$ for $\phi \in \Gamma$, so that

$$\sum_{i=1}^{n} |u(e_i)|^p \leq \pi_p(u)^p \int_{B'} \sum_{i=1}^{n} |\langle e_i,\xi\rangle|^p \nu(d\xi) \quad .$$

In particular we may take $n = 1$ to get

$$|u(x)|^p \leq \pi_p(u)^p \int_{B'} |\langle x,\xi\rangle|^p \nu(d\xi)$$

for all $x \in E$, the desired majorization of u. ■

Note that in the above construction $\|\nu\| = 1$ and $C = \pi_p(u)$, as promised. In general we can say that $\pi_p(u) = \inf\{C\|\nu\|\colon u \text{ admits an } L^p\text{-majorization with } C, \nu\}$.

It was necessary to take $p < +\infty$ in the above proof. The case $p = +\infty$ is not very interesting, since every u is $+\infty$-summing with $\pi_\infty(u) = \|u\|$. For this case we can take v as above. Then $\|v(x)\|_\infty = \|x\|$, so $\|u(x)\| \leq \pi_\infty(u)\|v(x)\|_\infty$. (There is, in general, no Pietsch measure for $p = +\infty$.)

These two Theorems will be referred to as the Pietsch factorization (for majorization) Theorem.

Lecture 3. Completely Summing Maps. Hilbert-Schmidt and Nuclear Maps

We can apply the Pietsch factorization theorem to give a second proof that if $q \geq p$ and u is p-summing, then u is q-summing. Indeed, we have

$$\begin{aligned} \|u(x)\| &\leq \pi_p(u)\Big[\int |\langle x,\xi\rangle|^p \nu(d\xi)\Big]^{1/p} \\ &\leq \pi_p(u)\Big[\int |\langle x,\xi\rangle|^q \nu(d\xi)\Big]^{1/q} \end{aligned}$$

so u has an L^q majorization as well, and we see that $\pi_q(u) \leq \pi_p(u)$.

We can extend the notion of p-summing map to *non-positive values of* p. Starting with finite sequences e, we require

$$\|u(e)\|_p \leq \pi_p(u)\|e\|_p^* .$$

<u>Note</u>: For $p = 0$, $\|x\|_0$ is the geometric mean. Thus our inequality reads

$$\left(\prod_{i=1}^{n} |u(e_i)|\right)^{1/n} \leq \pi_0(u) \sup_{|\xi|\leq 1} \left(\prod_{i=1}^{n} |\langle e_i,\xi\rangle|\right)^{1/n} .$$

(For $p < 0$ we have the same type of inequality. One should be alert to the fact that, since $p < 0$, the inequality reverses when we take p^{th} powers of both sides.) The Pietsch Theorems extend to this case, with the same proofs. E.g. for $p = 0$ we find a probability measure ν such that

$$\|u(x)\| \leq \pi_0(u) \sup_{|\xi|\leq 1} \exp\left(\int \log|\langle x,\xi\rangle|\nu(d\xi)\right) .$$

Using this approach one can see, even for $p \leq 0$, that p-summing maps are q-summing for all $q \geq p$. (The Hölder inequality approach does not work in this case, since the inequalities go the wrong way.)

But do there exist examples? After all, for $p < 0$, L^p makes no sense. The truth is that if $p \leq -1$ there are no p-summing maps (except in the trivial one-dimensional case). But there are interesting examples with $-1 < p \leq 0$. Accordingly we modify our previous definition and state that u is *completely summing* if and only if it is p-summing for all $p > -1$. One can show that every finite dimensional operator is completely summing.

We will eventually show that if u is p-summing for some $p < 1$ then it is completely summing, i.e. q-summing for all $q > -1$. The following summarizes the facts:

(1) If $\dim u(E) = 0$ or 1, then u is p-summing for $-\infty < p \leq +\infty$.

(2) If $2 \leq \dim u(E) < +\infty$, then the "cut" p_0 is at -1 and u is not (-1)-summing.

(3) If $\dim u(E) = +\infty$ then the cut is either ≥ 1 or $= -1$. In the latter case, u is not (-1)-summing.

The Pietsch Theorem gives us a subfactorization of any p-summing map u, $p > 0$:

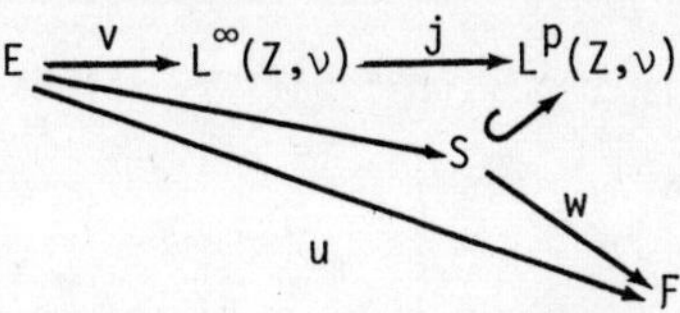

$\pi_p(u) = \min\|v\|\|w\|$ for all such subfactorizations of u.

If $p \geq 1$, we can also get an "over"-factorization (superfactorization) of u as follows. We are given $u: E \to F$, p-summing. Now $F \subseteq \ell^\infty(\Gamma)$ where Γ is the unit ball of F'. The above map $w: S \to F \subseteq \ell^\infty(\Gamma)$ can be extended to a map $\tilde{w}$ from $L^p(Z,\nu)$ into $\ell^\infty(\Gamma)$ by the Hahn-Banach Theorem. (In general, any space L^∞ is always "injective".) Thus we have the diagram:

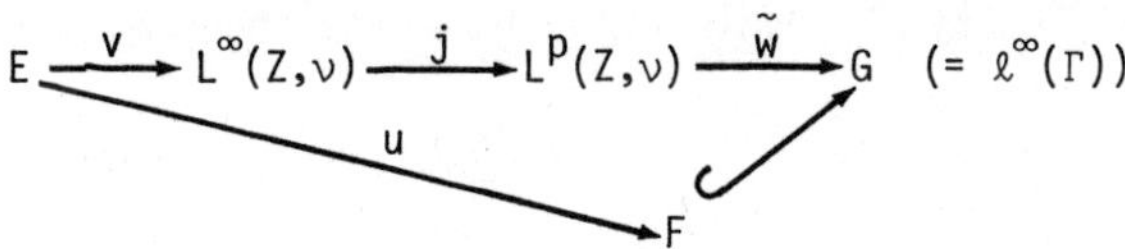

If μ is finite, we have $L^\infty(Z,\mu) \subset L^p(Z,\mu)$, and the embedding is p-summing. More generally, whether or not μ is finite, choose $\alpha \in L^p(Z,\mu)$ and define $(\alpha): L^\infty(Z,\mu) \to L^p(Z,\mu)$ to be multiplication by the function α.

<u>Proposition</u>: (α) is p-summing, and $\pi_p((\alpha)) = \|\alpha\|_{L^p}$.

<u>Proof</u>: We may assume $\|\alpha\|_{L^p} = 1$. Let $\nu = |\alpha|^p\mu$; then ν is a probability measure. We have the following factorization of (α):

$$\begin{array}{ccccc} L^\infty(Z,\mu) & \xrightarrow{\ id\ } & L^\infty(Z,|\alpha|^p\mu) & \xrightarrow{\ j\ } & L^p(Z,|\alpha|^p\mu) \\ & \searrow^{(\alpha)} & & & \downarrow a \\ & & & & L^p(Z,\mu) \end{array}$$

Here $a: L^p(Z,|\alpha|^p\mu) \to L^p(Z,\mu)$ is multiplication by α. Note that $\|a\| = 1$ here. The embedding j is p-summing, so we have that (α) is p-summing with norm at most 1. But it is obvious that $\pi_p((\alpha)) \geq \|\alpha\|_p = 1$, so in fact $\pi_p((\alpha)) = \|\alpha\|_p$. ∎

3.1. Additional Properties of p-summing Maps

(1) If $1 \leq p < +\infty$ and u is p-summing, then u is weakly compact.

Proof: Since a 1-summing map is p-summing for all $p > 1$, we may assume that $p > 1$. Then u factors through S, a closed subspace of L^p, hence a reflexive space. It follows that the image of the unit ball is relatively weakly compact, i.e. u is a weakly compact map. ¤

Note that u need not be compact. For example, the embedding of L^∞ in L^p is weakly compact but *not* compact.

(2) If $p < +\infty$, u transforms each weakly convergent sequence into a strongly convergent sequence.

Proof: Suppose x_n converges weakly to 0. We have

$$\|u(x_n)\| \leq \pi_p(u)\left[\int_Z |\langle x_n, \xi\rangle|^p \nu(d\xi)\right]^{1/p}$$

so that $\|u(x_n)\| \to 0$ by the dominated convergence theorem. ¤

Note that if E is *reflexive*, it follows that u is a *compact* operator.

(3) If $p < +\infty$, u transforms every weakly compact subset into a strongly compact subset.

Proof: This follows from (2) together with Eberlein's Theorem: A weakly compact set K is sequentially weakly compact, i.e. every sequence in K contains a weakly convergent subsequence. (Outline of the proof: We may reduce to the case of a separable space E. Then the unit ball B' of E' is metrizable and hence separable in the weak* topology. Hence there is a countable and weak* dense subset D' in E'. If $K \subset E$ is weakly compact, its topology coincides with the weak D' topology, which is metrizable.) ¤

(4) If E is finite dimensional, every linear map is completely summing. But if E is infinite dimensional, the identity I on E is never p-summing for $p < +\infty$. For if I is p-summing, the unit ball B of E is relatively weakly compact by (1), hence strongly compact by (3), and so E must be finite dimensional.

In other words, in every infinite dimensional Banach space E there exists a sequence which is scalarly ℓ^p but not ℓ^p. This fact is not at all easy to see without Pietsch's Theorem.

This is closely related to the Dvoretzky-Rogers Theorem: In any infinite dimensional Banach space E one can find a series of vectors which is unconditionally convergent but not absolutely convergent.

Sketch of the proof: Any 1-summing map $u: E \to F$ takes $s\ell^1(E)$ into $\ell^1(F)$. Let $uc(E)$ denote the unconditionally convergent sequences in E. (A sequence $\{e_n\}$ belongs to $uc(E)$ provided the series $\sum_n e_{\pi(n)}$ is convergent for all permutations π of the integers.) Obviously $uc(E) \subseteq s\ell^1(E)$, so if u is 1-summing it maps $uc(E)$ into $\ell^1(F)$. What about the converse?

Fact: The space $uc(E)$ is precisely the closure of the finite sequences in the $s\ell^1$ norm. In particular, it is a Banach space. Now suppose $u: E \to F$ takes $uc(E)$ into $\ell^1(F)$. By the closed graph theorem there is a constant C such that $\|u(e)\|_1 \leq C\|e\|_1^*$ for all $e \in uc(E)$. This holds in particular for *finite* sequences e, and so u is 1-summing.

Now if every unconditionally convergent series in E is absolutely convergent, we have $uc(E) = \ell^1(E)$, so the identity is 1-summing, and hence E is finite dimensional. ■

The Case of Hilbert Spaces

Recall the definition of a *Hilbert-Schmidt* map $u: E \to F$ between two Hilbert spaces: for some (and hence every) orthonormal basis (e_i) of E, the series $\sum_i \|u(e_i)\|^2$ is convergent. The sum is independent of the choice of basis. Moreover the adjoint u^* is also Hilbert-Schmidt, and if (f_j) is an orthonormal basis of F we have

$$\sum_i \|u(e_i)\|^2 = \sum_j \|u^*(f_j)\|^2$$

Indeed,

$$\sum_i \|u(e_i)\|^2 = \sum_{ij} |\langle ue_i, f_j\rangle|^2$$

$$= \sum_{ij} |\langle e_i, u^*f_j\rangle|^2$$

$$= \sum_j \|u^*f_j\|^2$$

We write

$$\|u\|_{HS} = [\sum_i \|u(e_i)\|^2]^{1/2} \quad .$$

This is a norm. It is almost trivial that $\|u\|_{HS} \geq \|u\|$. Moreover if u is Hilbert-Schmidt then u is compact. (For it is obviously the limit in norm of finite-rank operators.) It is also easy to see that the Hilbert-Schmidt operators form an ideal.

Note that $u \in HS(E,F)$, the space of Hilbert-Schmidt maps, if and only if $\sqrt{u^*u} \in HS(E)$. $\sqrt{u^*u}$ is compact and self-adjoint, with eigenvalues (λ_i). Since $u = v\sqrt{u^*u}$ where v is a partial isometry, we see that

$$\|u\|_{HS} = [\sum_i |\lambda_i|^2]^{1/2} \quad .$$

Next we recall Grothendieck's notion of *nuclear* maps. We say $u: E \to F$ is nuclear provided we can write

$$u = \sum_n \lambda_n e'_n \otimes f_n$$

(meaning $u(x) = \sum_n \lambda_n \langle e'_n, x\rangle f_n$) where $\{e'_n\}$ is a bounded sequence in E', $\{f_n\}$ is a bounded sequence in F, and $\sum_n |\lambda_n| < +\infty$. We define the nuclear norm $\|u\|_N = \inf \sum_n |\lambda_n|$, where the infimum is taken over all such representations of u in which $\|e'_n\|, \|f_n\| \leq 1$.

We have $HS(E,F) \hookrightarrow \mathcal{L}(E,F)$ where E, F are Hilbert. The injection is norm-decreasing, and $HS(E,F)$ is itself a Hilbert space.

The notion of nuclear map makes sense for arbitrary Banach spaces E, F. The space of nuclear maps $N(E,F) \hookrightarrow \mathcal{L}(E,F)$ is complete with respect to the nuclear norm. It is obvious that nuclear operators are compact.

If E and F are Hilbert spaces, then $u: E \to F$ is nuclear if and only if $\sqrt{u^*u}$ is nuclear, and $\|u\|_N = \|\sqrt{u^*u}\|_N$. In terms of the eigenvalues (λ_n) of $\sqrt{u^*u}$, we have the formula

$$\|u\|_N = \sum_n |\lambda_n| \quad .$$

Accordingly every nuclear map is Hilbert-Schmidt, but the converse is false. We have

$$\|u\|_N \geq \|u\|_{HS} \geq \|u\| \quad .$$

(The inequality $\|u\|_N \geq \|u\|$ is valid for arbitrary Banach spaces.)

Grothendieck proved that in Hilbert space an operator u is nuclear if and only if u is the product of two Hilbert-Schmidt maps. This is not difficult.

One may ask about the connections between the Hilbert-Schmidt property and 2-summing, and between the nuclear property and 1-summing. It turns out that Hilbert-Schmidt is the same as 2-summing, and in fact $\|u\|_{HS} = \pi_2(u)$. (A much deeper fact is that a Hilbert-Schmidt map is a p-summing for *all* p.)

In Banach spaces, nuclear maps are 1-summing, but the converse is false. For example the injection $L^\infty \to L^1$ is 1-summing but not nuclear (since it is not even compact). One has the inequality $\pi_1(u) \leq \|u\|_N$. Also, in Hilbert space, a Hilbert-Schmidt map need not be nuclear.

Lecture 4. p-integral Maps

The class of integral maps was introduced by Grothendieck. Our definition is somewhat different.

Definition: The map $u: E \to F$ is p-*integral* if there exists a factorization

$$E \to L^\infty(Z,\nu) \hookrightarrow L^p(Z,\nu) \to F \quad .$$

(The case $p = 1$ is close to Grothendieck's class.)

Similarly, we say that u is p-*nuclear* if it factors as

$$E \to \ell^\infty \xrightarrow{(\alpha)} \ell^p \to F$$

where (α) is multiplication by an element $\alpha \in \ell^p$.

4.1. Some Facts. (1) A p-nuclear map is p-integral. (Use (α) to define an appropriate probability measure.)

(2) The converse of (1) is false. However, Grothendieck proved that if either E or F is reflexive, then 1-integral implies nuclear. Note: Nuclear maps are always compact, while 1-integral maps need not be compact.

(3) If E is reflexive and $p > 1$, then p-integral implies p-nuclear.

(4) We can define "sub-p-integral" and "sub-p-nuclear" by introducing sub-factorizations in the obvious way. The following implications hold:

$$\begin{array}{ccc} \text{p-nuclear} & \Rightarrow & \text{p-integral} \\ \Downarrow & & \Downarrow \\ \text{sub p-nuclear} & & \text{sub p-integral} \\ & & \Updownarrow \\ & & \text{p-summing (Pietsch)} \end{array}$$

Note: We have norms for each class, e.g. for a p-nuclear map we take the infimum of $\|\alpha\|_p$ over all factorings (with $E \to \ell^\infty$ and $\ell^p \to F$ of norm ≤ 1).

(5) "Nuclear" is the same thing as "1-nuclear". For suppose that $u: E \to F$ is nuclear. Wrtie

$$u(x) = \sum_n \lambda_n \langle e_n', x \rangle f_n \quad \text{where} \quad \sum |\lambda_n| < \infty .$$

We can then factor u as follows:

$$E \xrightarrow{v} \ell^\infty \xrightarrow{(\lambda)} \ell^1 \xrightarrow{w} F$$

where $v(x) = (\langle e_n', x \rangle) \in \ell^\infty$, and $w((t_n)) = \sum_n t_n f_n \in F$; (λ) is multiplication by (λ_n). Thus u is 1-nuclear; and conversely.

(6) In Hilbert spaces, the following properties are equivalent: (a) Hilbert-Schmidt, (b) 2-integral, (c) 2-nuclear, (d) 2-sub-integral, (e) 2-subnuclear, (f) 2-summing.

Proof: In view of (4) above, it is enough to show 2-summing ⇒ Hilbert-Schmidt ⇒ 2-nuclear ⇒ 2-summing.

Now u 2-nuclear ⇒ u factors through $\ell^\infty \overset{\alpha}{\hookrightarrow} \ell^2$ ⇒ u is 2-summing with $\pi_2(u) \leq N_2(u)$.

That 2-summing implies Hilbert-Schmidt is trivial. Indeed, consider an orthonormal basis $(e_n) = e$ in E. It is scalarly ℓ^2 with $\|e\|_2^* = 1$. If u is 2-summing, $u(e_n)$ is then an ℓ^2 sequence, so

$$[\sum|u(e_n)|^2]^{1/2} \leq \pi_2(u)\|e\|_2^* = \pi_2(u) \quad .$$

Thus u is Hilbert-Schmidt with $\|u\|_{HS} \leq \pi_2(u)$.

Finally, we show that u Hilbert-Schmidt ⇒ u is 2-nuclear, with $N_2(u) \leq \|u\|_{HS}$. Recall that if e, f are orthonormal bases in E, F then

$$[\sum|u(e_i)|^2]^{1/2} = \|u\|_{HS} = [\sum|u^*(f_j)|^2]^{1/2} \quad .$$

Define $v: E \to \ell^\infty$ by

$$v(x) = (\langle x, u^*f_n\rangle/\|u^*f_n\|) \quad .$$

Note that if $\alpha_n = \|u^*f_n\|$ then $\alpha \in \ell^2$ with $\|\alpha\|_2 = \|u\|_{HS}$. Finally define $w: \ell^2 \to F$ by $w(t) = \sum_n t_n f_n$. Then, for $x \in E$,

$$u(x) = \sum_n (u(x), f_n)f_n = \sum_n \frac{\langle x, u^*f_n\rangle}{\|u^*f_n\|} \cdot \|u^*f_n\| \cdot f_n \quad .$$

Clearly this shows that $u = w\circ(\alpha)\circ v$. Since w and v have norms ≤ 1, we have $N_2(u) \leq \|\alpha\|_2 = \|u\|_{HS}$.

II. Cylindrical Probabilities and Radonifying Maps

First recall what is meant by a *Radon measure* μ on a Hausdorff space X. We suppose that μ is a Borel probability measure, and in addition that μ is *tight* or *inner regular*: $\mu(B) = \sup_{K \subset B} \mu(K)$, K compact. As a consequence,

$$\mu(B) = \inf_{\mathcal{O} \supset B} \mu(\mathcal{O}) \ , \quad \mathcal{O} \text{ open.}$$

This tightness property implies some rather general limit properties. Thus, if $\{\mathcal{O}_i\}_{i \in I}$ is an increasing system of open sets, then $\lim_i \mu(\mathcal{O}_i) = \mu(\bigcup_{i \in I} \mathcal{O}_i)$. A similar result holds for decreasing systems of closed sets.

Now let Y be another topological space and let $u: X \to Y$. Let μ be a Borel measure on X. If u is continuous we can define the image Borel measure $u\mu$ on Y by $(u\mu)(B) = \mu(u^{-1}B)$. More generally, we say that u is μ-*Lusin measurable* provided that, for all $\varepsilon > 0$, there is a compact $K_\varepsilon \subset X$ such that $u|K_\varepsilon$ is continuous and $\mu(X \setminus K_\varepsilon) < \varepsilon$. Then if μ is a Radon measure it follows that $u\mu$ is Radon, and moreover the formula $u\mu(B) = \mu(u^{-1}B)$ holds for all measurable sets B, not merely Borel sets. Moreover, for $B \subset Y$, $u^{-1}(B)$ is μ-measurable if and only if B is $u\mu$-measurable. (There is no "image measure catastrophe".)

Suppose that μ is a Radon probability on a Hilbert space E. If $F \subseteq E$ is a finite dimensional subspace let P_F be orthogonal projection onto F. Then $P_F(\mu) = \mu_F$ is a Radon probability on F. The system of measures μ_F has the following coherence property: If $G \subseteq F \subseteq E$ and $P_{G,F}$ is orthogonal projection from F onto G, then $P_G = P_{GF}P_F$, so that

$$\mu_G = P_{GF}(\mu_F) \quad .$$

Moreover, since μ is a Radon measure, it is determined by the system of all μ_F's. It is enough to recover $\mu(K)$ for compact K. Now the sets $P_F^{-1}(p_F K)$ form a decreasing system of closed sets, and $K = \bigcap_F P_F^{-1}(P_F K)$. Accordingly,

$$\mu(K) = \inf_F \mu(P_F^{-1}(P_F K))$$
$$= \inf_F (P_F\mu)(P_F K)$$
$$= \inf_F \mu_F(K_F) \quad , \quad \text{where} \quad K_F = P_F(K) \ .$$

Note: For *separable* E, every Borel probability μ is Radon.

Definition: A *cylindrical probability* on a Hilbert space E is a system of Radon probabilities μ_F on the finite dimensional subspaces F of E, satisfying the coherence condition $\mu_G = P_{G,F}(\mu_F)$ if $G \subseteq F$.

These fit together to define a *finitely* additive measure on the ring of cylinder sets (Borel sets of the form $P_F^{-1}(B)$, $B \subseteq F$). *Prokhorov* gave a necessary and sufficient condition for a cylindrical probability to be a Radon probability, i.e. to be countably additive.

4.2. Theorem (Prokhorov): Let $(\mu_F)_{F \in \mathcal{F}}$ be a cylindrical probability on E. The following condition is necessary and sufficient for (μ_F) to be derived from a Radon probability μ: For each $\varepsilon > 0$, there is a compact $K \subseteq E$ such that, for all $F \in \mathcal{F}$, $\mu_F(K_F) \geq 1 - \varepsilon$.

Remarks: The necessity is trivial. Indeed, if μ is Radon, we can choose K so that $\mu(K) \geq 1 - \varepsilon$. Then, for any F,

$$\mu_F(K_F) = \mu(P_F^{-1}(P_F K)) \geq \mu(K) \geq 1 - \varepsilon \ .$$

The sufficiency of Prokhorov's condition is more remarkable, but we will not give the proof here. (See the References.)

Gaussian Measures

For E an n-dimensional Hilbert space, the Gauss law Γ_E is $\frac{1}{(\sqrt{2\pi})^n} e^{-|x|^2/2} dx$. (For $n = 0$ we take the δ-measure on $\{0\}$.) Under orthogonal projection $E \to F$, Γ_E goes into Γ_F. This is obvious because Γ_E factors as $\Gamma_F \otimes \Gamma_{F^\perp}$.

Now let E be an infinite dimensional Hilbert space, $\mathcal{F}$ the family of finite dimensional subspaces. Then $\{\Gamma_F : F \in \mathcal{F}\}$ is a cylindrical probability on E. But it is *not* a Radon probability, i.e. it is not countably additive. Indeed, it is easy to see that Prokhorov's condition is not satisfied. This is related to the fact that the measure of the n-dimensional ball of radius R (R fixed) tends to 0 as $n \to \infty$. So if $K \subseteq E$ is compact, $K \subseteq$ some B_R, and so $\Gamma_F(B_R) \supseteq \Gamma_F(K)$; thus $\Gamma_F(K) \to 0$ as F increases. (Note that in n dimensions $\Gamma_F(B_R) \leq \left(\frac{1}{\sqrt{2\pi}}\int_{-R}^{R} e^{-x^2/2}dx\right)^n$, and this tends to 0 as $n \to \infty$.)

Lecture 5. Completely Summing Maps: Six Equivalent Properties. p-Radonifying Maps

Suppose that μ is a cylindrical probability on E and that $u: E \to F$ is a continuous linear map. If G is a finite dimensional subspace of F there exists a finite dimensional subspace $H \subseteq E$ such that the following diagram commutes:

$$\begin{array}{ccc} E & \xrightarrow{u} & F \\ {\scriptstyle P_H}\downarrow & & \downarrow{\scriptstyle P_G} \\ H & \xrightarrow{v} & G \end{array}$$

We then define $(u\mu)_G = v(\mu_H)$. This defines a cylindrical probability $u\mu$ on F.

5.1. Theorem: Let E, F be Hilbert spaces, $u: E \to F$ continuous and linear. The following six properties are equivalent:

(1) u is completely summing.

(2) u is p-summing for some $p < +\infty$.

(3) $u\Gamma$ is a Radon measure (Γ = Gauss law on E).

(4) u^* is completely summing.

(5) u^* is p-summing for some $p < +\infty$.

(6) $u^*\Gamma^*$ is a Radon measure.

Proof: (1) ⇒ (2) and (4) ⇒ (5) are trivial. The important steps are (2) ⇒ (3) and (3) ⇒ (4) together with their analogues (5) ⇒ (6) and (6) ⇒ (1). The proof that (2) ⇒ (3) takes a lot of work, and we have to develop some related material. For this purpose it is useful to go outside the category of Hilbert spaces. (We will continue the proof of the theorem in Lecture 6.)

Accordingly, let E be a Hausdorff locally convex topological vector space. In place of orthogonal projections we have the factor spaces of E: $F = E/N$ where N is a closed subspace of finite codimension. Write $P_F: E \to E/N = F$ for the quotient map. If μ is a Radon measure on E we have $\mu_F = P_F\mu$ on each finite-dimensional factor space F. Now suppose that $M \subseteq N \subseteq E$ with $E/M = F$, $E/N = G$. Since $E/N = (E/M)/(N/M)$ we have a natural map $P_{G,F}: F \to G$; it is the unique map such that the following diagram commutes:

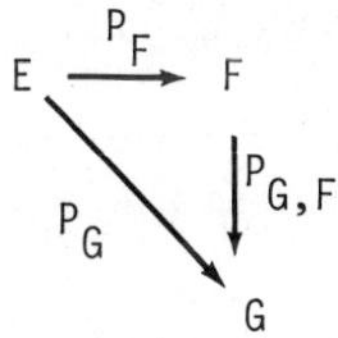

Then we have the "coherence property"

$$\mu_G = P_{GF}\mu_F \ .$$

Conversely, we take this as the definition of cylindrical measures in the category of topological vector spaces. (Local convexity is needed to guarantee the existence of enough finite dimensional factor spaces.)

Of course, we use only the duality (E,E'), not the full topology of E. Only the weak topology plays a role. So, more generally, we can define a cylinder measure on a given pair of spaces in duality; e.g. we could take the pair (L^∞,L^1) even though L^1 is not the full dual of L^∞.

But note that the strong topology of E enters the picture when we ask whether a cylindrical measure on E comes from a *Radon* measure. In any case there is at most one such measure. For if $K \subset E$ is compact, then, as with Hilbert spaces, $K = \bigcap_F P_F^{-1}(P_F K)$, so that we must have $\mu(K) = \inf_F \mu_F(K_F)$. Also we observe that Prokhorov's Theorem generalizes: the criterion for (μ_F) to be Radon is that for each $\varepsilon > 0$ there exists a compact K such that for all finite dimensional factor spaces F, $\mu_F(K_F) \geq 1 - \varepsilon$.

Now consider a dual pair (E,E') and $u: E \to F$ where F is finite dimensional. Let μ be a cylindrical probability on E. Then it induces a Radon probability μ_u on F. For u factors through the finite dimensional quotient space E/N (N = nullspace of u), and $\mu_{E/N}$ pushes forward to F. This leads to a *new definition of a cylindrical measure*: a family (μ_u) indexed by the class of all continuous linear maps $u: E \to F$ with F finite dimensional. The coherence condition says that if we have

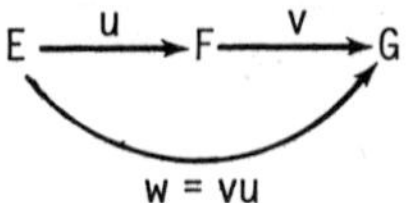

then $\mu_w = v(\mu_u)$.

To avoid the logical objection that the above class of u's is too large to be a set, we could take concrete spaces $\mathbb{R}^n$ for the F's. Then our definition says: for each finite set $\xi_1,\xi_2,\ldots,\xi_n \in E'$ there is a measure $\mu_{\xi_1\xi_2\ldots\xi_n}$ on $\mathbb{R}^n$, such that if (α_{ij}) is a matrix and $\eta_i = \sum_{j=1}^{n} \alpha_{ij}\xi_j$, $1 \leq i \leq m$, then

$$\mu_{\eta_1,\ldots,\eta_m} = (\alpha)\cdot\mu_{\xi_1,\ldots,\xi_n} .$$

Let μ be such a cylindrical probability on E, and let $u: E \to F$, F arbitrary, u weakly continuous. Then for $v: F \to G$, with G finite-dimensional, we have $vu: E \to G$, so we define $u(\mu)$ by

$$[u(\mu)]_v = \mu_{vu} .$$

It is easy to verify that this is a cylindrical probability on F.

<u>An interesting example</u>: Let E be a non-reflexive Banach space; $E \subseteq E''$. Let a'' be a point in $E''\backslash E$. Then the Dirac $\delta_{a''}$ measure on E'' is Radon, but it defines a *non-Radon* cylindrical measure on E:

$$\begin{array}{ccl} E & \xrightarrow{u} & F = F'' \\ \downarrow & \nearrow_{u''} & \\ E'' & & \end{array} \qquad (\delta_{a''})_u \text{ is defined to be } \delta_{u''(a'')}$$

Let E be a Banach space, and suppose that μ is a Radon measure on E. We then define, for $-\infty < p \leq +\infty$,

$$\|\mu\|_p = \left(\int_E \|x\|^p \mu(dx)\right)^{1/p}$$

with the usual conventions for $p = \infty$ or 0. (Thus, $\|\mu\|_0 = \exp\int_E \log\|x\|\mu(dx)$ and $\|\mu\|_\infty = \text{ess sup}(\mu)\|x\|$.) We say that μ is of *order* p if $\|\mu\|_p < \infty$. Similarly, we define the *scalar order* by setting

$$\|\mu\|_p^* = \sup_{|\xi|\leq 1} \|\xi\cdot\mu\|_p$$

where $\xi \in E'$, so that $\xi\cdot\mu$ is a measure on $\mathbb{R}$. Note that this makes sense for cylindrical μ.

Observe that the case of sequences $\|e\|_p$, $\|e\|_p^*$ is included in the above if we associate to a sequence an appropriate discrete measure. We can thus pass from the notion of "p-summing" to "p-Radonifying".

Definition: $u: E \to F$ is p-*Radonifying* provided that if μ is a cylindrical measure on E of scalar order p, then $u(\mu)$ is a Radon measure on F of true order p. (Clearly this says in particular that u is p-summing.) By analogy with p-summing maps, it would be desirable if in addition we had an inequality of the form

$$(*) \qquad \|u(\mu)\|_p \leq \pi_p(u)\|\mu\|_p^*.$$

Unfortunately, this need not be true. Of course, conversely, if u is p-Radonifying and (*) holds, then u is p-summing with the same constant. (If we don't assume (*), one might hope to get a constant by some sort of Banach-Steinhaus argument. But this breaks down; for example, the μ's don't form a vector space.)

In detail: suppose that (e_n) is a sequence in E. Let $c_n > 0$, $\sum_n c_n = 1$. Consider the Radon probability

$$\mu = \sum c_n \delta_{(c_n^{-1/p} e_n)} \quad .$$

We have

$$\|\mu\|_p = \left(\sum_n c_n \|c_n^{-1/p} e_n\|^p\right)^{1/p}$$

$$= \left(\sum_n \|e_n\|^p\right)^{1/p} = \|e\|_p \quad .$$

Similarly $\|\mu\|_p^* = \|e\|_p^*$. Also $u(\mu)$ is associated in the same way to $u(e)$. So, even without (*), we see that if u is p-Radonifying it is p-summing.

For the converse we present a partial result; the general case is still open.

Suppose that u is p-summing and that μ is a finite system of discrete masses. Then we have $\|u\mu\|_p \leq \pi_p(u)\|\mu\|_p^*$.

One would like to pass somehow to more general measures μ. This has not been completely achieved. The basic idea is to *approximate* a given μ by finite, discrete μ_j's, for which we have $\|u\mu_j\|_p \leq \pi_p(u)\|\mu_j\|_p^*$.

More precisely, we need a topology on the space of measures with a number of properties. First of all, we want to be able to approximate an arbitrary μ by finitely supported discrete measures μ_j such that $\|\mu_j\|_p^* \leq \|\mu\|_p^*$. Then we will have $\|u\mu_j\|_p \leq \pi_p(u)\|\mu\|_p^*$. Furthermore, the approximation process must be such that when $\mu_j \to \mu$ we also have $u\mu_j \to u\mu$. If so, we have $u\mu_j \to u\mu$ with $\|u\mu_j\|_p \leq$ constant $< \infty$. Then we will need a theorem to the effect that the set of *Radon measures* whose p-order is $\leq M$ is closed in the space of cylindrical probabilites in our topology. If all this is true, we will end up with $u(\mu)$ Radon such that $\|u(\mu)\|_p \leq \pi_p(u)\|\mu\|_p^*$.

We introduce the *strict*("etroite") *topology*: $\mu_j \to \mu$ if and only if, for every bounded continuous ϕ on E, we have

(*) $$\mu_j(\phi) \to \mu(\phi) \quad .$$

For locally compact E this is stronger than vague (= weak*) convergence, which only requires (*) for ϕ continuous with compact support. For example, on $\mathbb{R}$, $\delta_{(n)} \to 0$ vaguely but not strictly. But if probability measures μ_j converge vaguely to a *probability* measure μ, so that no mass is "lost", then $\mu_j \to \mu$ strictly.

We can extend this notion to cylindrical probabilities. We say $\mu_j \to \mu$ provided that, for all $u\colon E \to F$ with finite dimensional range, we have $u(\mu_j) \to u(\mu)$ strictly. (The topology thus defined is called the *cylindrical topology*.) It is then trivial that, if $v\colon E \to G$ is continuous and linear, then $v(\mu_j) \to v(\mu)$. So at least one of our desiderata is met.

Lecture 6. Radonification Theorem

Let E, F be Banach spaces, with $u\colon E \to F$ a p-summing map. Let μ be a cylindrical probability on E of scalar order p: $\|\mu\|_p^* < \infty$. Is $u(\mu)$ of true order p with $\|u(\mu)\|_p \leq \pi_p(u)\|\mu\|_p^*$? We know that this is so if μ is a finite sum of point masses.

As indicated above, we equip the space of cylindrical probabilities with the weakest topology which is "strict" on the finite dimensional projections. Equivalently, $\mu_j \to \mu$ provided that $\int \phi d\mu_j \to \int \phi d\mu$ for all bounded, continuous cylinder functions ϕ.

Suppose that there exist finite discrete measures μ_j, such that $\mu_j \to \mu$ in the cylindrical topology, with $\|\mu_j\|_p^* \leq \|\mu\|_p^*$ for all j. We then say that μ is *approximable of scalar order* $\|\mu\|_p^*$.

It is automatic that $u(\mu_j) \to u(\mu)$ cylindrically. Moreover $\|u(\mu_j)\|_p \leq \pi_p(u)\|\mu_j\|_p^* \leq \pi_p(u)\|\mu\|_p^*$. From this we would like to deduce that $u(\mu)$ is Radon with $\|u(\mu)\|_p \leq \pi_p(u)\|\mu\|_p^*$.

We will try to establish the following (with some additional conditions on F):

The set S of Radon measures ν on F with $\|\nu\|_p \leq M$ is compact in the *strict* topology. Then a fortiori it is compact and hence closed in the cylindrical topology.

(1) The set S is closed. First, the map $\nu \mapsto \nu(\phi)$ is continuous for $\phi \in BC(F)$. If ϕ is continuous and ≥ 0 but not necessarily bounded, then the map $\nu \mapsto \nu(\phi)$ is lower semicontinuous (since it is the supremum of continuous maps). In particular, taking $\phi(y) = \|y\|^p$, it follows that $\nu \mapsto \|\nu\|_p$ is lower semicontinuous, so that S is closed.

(2) S is relatively compact. To see this we appeal to a theorem of Prokhorov and Le Cam:

6.1. Theorem: Let P be the probability measures on X with the strict topology. Let $M \subseteq P$. Suppose that for every $\varepsilon > 0$ there is a compact $K \subseteq X$ such that, for all $\nu \in M$, $\nu(K) \geq 1 - \varepsilon$. Then M is relatively compact in the strict topology.

To apply this condition in our case, we use Chebyshev's inequality: if $\|\nu\|_p \leq M$ and $B(R)$ is the ball of radius R in F, then

$$\nu(F \setminus B(R)) \leq M^p/R^p.$$

Unfortunately, B(R) isn't compact when F is infinite dimensional; if it were we could apply the above theorem. To get round this difficulty, we pass from F to the second dual F": $E \overset{u}{\to} F \to F''$. The R-ball B(R) in F" is *weak* compact*. Moreover the function $\|y\|^p$ is lower semi-continuous in the weak* topology, and therefore the p-order of a Radon probability relative to $\sigma(F'',F')$ is well-defined.

Conclusion: Suppose that μ is a cylindrical probability on E, approximable of scalar order $\|\mu\|_p^*$. Then $u(\mu)$ is Radon on $\sigma(F'',F')$, and $\|u(\mu)\|_p \leq \pi_p(u)\|\mu\|_p^*$.

We are left with two problems:

(1) Is every μ on E approximable?

(2) Can we return to F from F"? (Here there are counterexamples.)

These problems can be overcome under various additional hypotheses. At present it is not known whether (1) is always possible. It is worth noting that (1) is related to Banach's *metric approximation property*. A Banach space X has the latter property provided there exists a net π_j of finite rank operators on X with $\|\pi_j\| \leq 1$ and $\pi_j \to 1$ pointwise. Banach conjectured that every Banach space has this property. But Enflo (1972) found counterexamples; in fact there are "bad" subspaces of L^p if $p > 2$. (Recently Szankowski proved that the space B(H) of bounded operators on Hilbert space does not have the metric approximation property.)

6.2. Theorem: If E' has the metric approximation property then every cylindrical probability μ on E of scalar order p is approximable. (Note: The hypothesis implies that E has the metric approximation property as well.)

As an application we have the following result.

6.3. Theorem: Suppose that $p \geq 1$. Then, with no additional hypotheses on E, it follows that $u(\mu)$ is a Radon measure on F'' with $\|u(\mu)\|_p \leq \pi_p(u)\|\mu\|_p^*$.

Proof: Consider the Pietsch factorization of u:

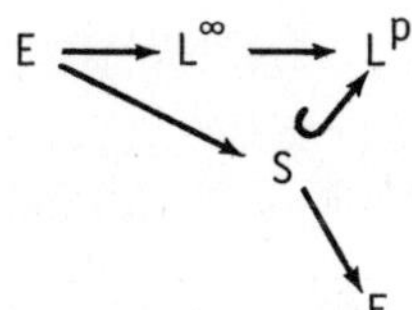

It is known that L^∞ and its dual posses the metric approximation property (all "classical" Banach spaces have this property). Therefore the image of μ on L^∞ is approximable. Since the canonical injection $L^\infty \to L^p$ is p-summing, the image of μ will be Radon of true order p on $\sigma(L^{p''},L^{p'})$, and hence on $\sigma(F'',F')$. ∎

Note that this argument fails if $p < 1$, since then L^p is not locally convex and there are no cylindrical probabilities except δ_0.

Next we have the "bidual" problem: F'' versus F.

(1) If F is reflexive, $u(\mu)$ is a Radon measure on $\sigma(F,F')$. Then, by a theorem of Phillips, this measure extends uniquely to a Radon measure on F.

(2) If $1 < p < +\infty$, we can use Pietsch factorization together with the fact that L^p is reflexive to end up with a Radon measure on F; we need not introduce F''.

(3) If $p = \infty$ and F is not reflexive, the theorem is false; there always exists a non-Radon measure of scalar order ∞. (Take $\delta_{a''}$, $a'' \in F''\backslash F$.)

However if $p = 1$ it is sufficient that either E' or F has the *Radon-Nikodym property* (R.N.P.): A Banach space X has the R.N.P. provided that every X-valued measure bounded by a positive measure has a density. Every reflexive space has the R.N.P.; likewise, every separable dual space.

6.4. Theorem (Pietsch): F has the R.N.P. if and only if every 1-summing operator $E \to F$ is 1-Radonifying.

Summary

6.5. Theorem of Radonification: The "perfect case" is $1 < p < +\infty$.

In general we get $u(\mu)$ on F'' rather than F. For $p = +\infty$ we require that F be reflexive. For $p = +1$ we require that either E' or F has the R.N.P. For $p < +1$, reflexivity of F eliminates the bidual problem, but we seem to need the metric approximation property for E', although this may be simply an artifact of the method of proof.

Now we return to the theorem of "six equivalent properties" stated earlier (Theorem 5.1). Thus let E, F be Hilbert spaces, $u: E \to F$.

(2) ⇒ (3): Suppose that u is p-summing, $p < +\infty$. Let Γ be Gauss measure on E. We have to show that $u\Gamma$ is Radon. Now we have

$$\|\Gamma\|_p^* = \sup_{|\xi|\leq 1} \|\xi(\Gamma)\|_p \quad .$$

Here $\xi(\Gamma)$ is of course a normal law on $\mathbb{R}$, with variance related to $\|\xi\|$. One computes $\|\xi(\Gamma)\|_p = \|\xi\|\|\gamma\|_p$, where γ = Gauss law on $\mathbb{R}$. Thus $\|\Gamma\|_p^* = \|\gamma\|_p$. So, by the "Theorem of Radonification" discussed above, $u(\Gamma)$ is Radon of order p (since $p < \infty$).

We sketch the remainder (details to be supplied in Lecture 7).

(3) ⇒ (4): Assume that $u(\Gamma)$ is Radon: We have to show that u^* is p-summing, for all p. This can be done using Pietsch factorization, with $u(\Gamma)$ the Pietsch measure on F.

(4) ⇒ (5): Trivial. (5) ⇒ (6): $u^*(\Gamma^*)$ is Radon by the argument that (2) ⇒ (1). Then (6) ⇒ (1) by the argument that (3) ⇒ (4).

A further equivalent property is:

(7) u is Hilbert-Schmidt.

This is obvious, since we already know that Hilbert-Schmidt is equivalent to 2-summing.

Lecture 7. p-Gauss Laws

(3) ⇒ (4): We must prove: if $u: E \to F$ with $u(\Gamma)$ Radon on $\sigma(F'',F')$, then u^* is completely summing.

Note: We really need only assume that E is a Hilbert space, so that Γ is defined; F can be any Banach space.

We also remark that u p-summing need not imply that u^* is p-summing; however Pietsch has shown that u^{**} must be p-summing--a difficult theorem.

By the fundamental property of the Gauss law, if $\xi \in E$ then $\xi(\Gamma)$ is the normal law on $\mathbb{R}$ given by $\|\xi\|\cdot\gamma$ (here γ is Gauss law on $\mathbb{R}$ and $\|\xi\|$ stands for the homothecy, multiplication by $\|\xi\|$). Accordingly

$$\|\xi(\Gamma)\|_p = \|\xi\| \ \|\gamma\|_p$$

so that

$$\|\xi\| = \|\xi(\Gamma)\|_p/\|\gamma\|_p \quad .$$

Now apply this to $\xi = u^*\eta$, with $\eta \in F'$. We have

$$\begin{aligned} |u^*\eta| &= \|(u^*\eta)(\Gamma)\|_p/\|\gamma\|_p \\ &= \|\eta(u(\Gamma))\|_p/\|\gamma\|_p \quad . \end{aligned}$$

Because $u(\Gamma)$ is a *Radon* probability measure, we get

$$|u^*\eta| = \frac{1}{\|\gamma\|_p}\left(\int_{\mathbb{R}} |t|^p (\eta(u\Gamma))(dt)\right)^{1/p} = \frac{1}{\|\gamma\|_p}\left(\int_{F''} |\langle \eta,y\rangle|^p (u\Gamma)(dy)\right)^{1/p}$$

by the change-of-variables formula. If we write $\|u\Gamma\|_p = \alpha$, then, assuming α is finite, we have

$$|u^*\eta| = \frac{\alpha}{\|\gamma\|_p}\left(\int_{F''} \frac{|\langle \eta,y\rangle|^p}{|y|^p}\, \alpha^{-p}|y|^p (u\Gamma)(dy)\right)^{1/p} \quad .$$

Now consider $Z = \sigma(F'',F')$ and note that $\nu(dy) = \alpha^{-p}\, y^{p}(u\Gamma)(dy)$ is a probability measure on Z. To get a Pietsch factorization of u^* we define $v\colon F' \to L^{\infty}(Z,\nu)$ by

$$v(\eta)(y) = \frac{\langle \eta, y\rangle}{|y|} \quad .$$

Then $\|v(\eta)\|_{\infty} \leq |\eta|$ so v is continuous and linear with $\|v\| \leq 1$. Now we have

$$|u^*\eta| \leq \frac{\alpha}{\|\gamma\|_p} \|v(\eta)\|_{L^p(Z,\nu)} \quad ,$$

so, by Pietsch majorization, u^* is p-summing with $\pi_p(u^*) \leq \|u\Gamma\|_p / \|\gamma\|_p$.

Of course, we have been assuming all along that $\|u\Gamma\|_p$ is finite. To prove this is a serious difficulty, but there is a result of Shepp, Landau, and Fernique which settles it:

<u>Theorem</u>: If $u\Gamma$ is a "Gaussian process" then $\|u\Gamma\|_p < +\infty$ for all $p < +\infty$.

A second approach is to consider $p < 0$. Then we always have $\|\mu\|_p < \infty$. For $|y|^p > 0$ implies $\int |y|^p \mu(dy) > 0$ (possibly $+\infty$). (Note that $\int |y|^p \mu(dy) = 0$ if and only if $\mu = \delta$. But this, for $\mu = u\Gamma$, implies that $\eta(u\Gamma) = \delta$ for every $\eta \in F'$; that is, $u^*\eta(\Gamma) = \delta$, so $u^*\eta = 0$, hence $u = 0$, a trivial case which we exclude.) Then, since p is negative, $\left(\int |y|^p \mu(dy)\right)^{1/p}$ is finite or 0, but never $+\infty$. However, if $p \leq -1$ then $\|u\Gamma\|_p = 0 = \|\gamma\|_p$ so their ratio is indeterminate. But if $-1 < p < 0$ then $\|\gamma\|_p \neq 0$ or ∞ and we get $\pi_p(u^*) \leq \|u\Gamma\|_p / \|\gamma\|_p$.

Now (cf. Lecture 3) it follows that u^* is completely summing, i.e. p-summing for all $p > -1$. This finishes the proof of Theorem 5.1. ∎

<u>Remark</u>: Oddly enough, in the above argument it was easier to treat the case of negative p, though in general it is more difficult. For, the smaller p, the smaller $\|u\Gamma\|_p$, and so the easier it is to show that u^* is p-summing. (Incidentally, the Shepp-Landau-Fernique result does *not* work for $p < 0$.)

Now our earlier estimate says that if u is p-summing then

$$\|u\Gamma\|_p \leq \pi_p(u)\|\gamma\|_p \quad .$$

Accordingly

$$\pi_p(u) \geq \frac{\|u\Gamma\|_p}{\|\gamma\|_p} \geq \pi_p(u^*) \quad .$$

Thus we have:

7.1. <u>Proposition</u>: In the case of Hilbert space operators,

$$\pi_p(u) = \pi_p(u^*) = \frac{\|u\Gamma\|_p}{\|\gamma\|_p} = \frac{\|u^*\Gamma^*\|_p}{\|\gamma\|_p} \ .$$

<u>Note</u>: Of course we can consider more general cylindrical probabilities μ. We then have

$$\pi_p(u) \geq \|u\mu\|_p / \|\mu\|_p^*$$

with equality if $\mu = \Gamma$.

Suppose, again, $u: E \to F$, with E, F Hilbert spaces. Suppose that $p \leq q$. Then we have

$$\pi_p(u)\|\gamma\|_p/\|\gamma\|_q \leq \pi_q(u) \leq \pi_p(u) \ .$$

For

$$\pi_q(u) = \frac{\|u\Gamma\|_q}{\|\gamma\|_q} \geq \frac{\|u\Gamma\|_p}{\|\gamma\|_q} = \pi_p(u)\frac{\|\gamma\|_p}{\|\gamma\|_q} \ .$$

Thus we have $\pi_p(u) \leq C_{pq}\pi_q(u)$ where $C_{pq} \leq \|\gamma\|_q/\|\gamma\|_p$ is a universal constant (independent of u). Note that $C_{pq} \to \infty$ as $p \to -1$ since $\|\gamma\|_p \to 0$.

<u>The "p-Gauss" Laws of P. Lévy</u>

$\gamma_2(dx) = \frac{1}{\sqrt{2\pi}} e^{-x^2/2}dx$ has Fourier transform $e^{-\xi^2/2}$.

Now consider the function $e^{-|\xi|^p}$, $0 < p \leq 2$. This is a positive definite function, so by Bochner's Theorem it is the Fourier transform of a probability measure which we call γ_p, the p-*Gauss law*. For $p = 1$ we have the Cauchy law $\frac{1}{\pi} \cdot \frac{dx}{1+x^2}$. In general one can show that γ_p and Lebesgue measure dx are mutually absolutely continuous, and the density of γ_p is asymptotic to constant $\times |x|^{-(p+1)}$ if $p < 2$. (Of course, for $p = 2$, γ_2 has a rapidly decreasing density, but this is false for smaller values of p.)

Thus, for $p < 2$, γ_p has q-moments only for $q < p$. It is convenient to introduce the notation $\bar{p} = \begin{cases} p \text{ if } p < 2 \\ \infty \text{ if } p = 2 \end{cases}$. Then $\|\gamma_p\|_q < \infty$ for $q < \bar{p}$.

Let $\tau \cdot \gamma_p$ denote the dilation of γ_p by τ. Its Fourier transform is $e^{-|\tau\xi|^p}$. Now consider the convolution

$$(\tau_1\gamma_p)*(\tau_2\gamma_p)*\cdots*(\tau_n\gamma_p) \quad .$$

This has Fourier transform

$$e^{-|\tau_1\xi|^p} e^{-|\tau_2\xi|^p} \cdots e^{-|\tau_n\xi|^p} = e^{-|\tau\xi|^p}$$

where $\tau = (|\tau_1|^p+|\tau_2|^p+\cdots+|\tau_n|^p)^{1/p}$. So the above convolution is $\tau\cdot\gamma_p$. This is the law for the sum $\tau_1 X_1+\cdots+\tau_n X_n$ where the X_j are independent R.V.'s with law γ_p. (For $p = 2$ we have a well-known result about Gaussian R.V.'s.)

Lecture 8. Proof of the Pietsch Conjecture

We shall construct "p-Gauss" laws Γ_p on certain Banach spaces. To this end it is necessary to introduce some Fourier transform machinery.

If μ is a Radon measure on E, we define the Fourier transform $\hat{\mu}$ on E' by

$$\hat{\mu}(\xi) = \int_E e^{-i\langle \xi,x\rangle}\mu(dx) \quad .$$

Now suppose $u: E \to F$ is a continuous linear map. Then $u(\mu)$ is a Radon measure on F, and we have

$$\begin{aligned}(u\mu)\hat{}\,(\eta) &= \int_F e^{-i\langle \eta,y\rangle}(u\mu)(dy) \\ &= \int_E e^{-i\langle \eta,ux\rangle}\mu(dx) \\ &= \int_E e^{-i\langle {}^t u\eta,x\rangle}\mu(dx) = \hat{\mu}({}^t u\eta) \quad .\end{aligned}$$

That is, $(u\mu)\hat{} = \hat{\mu}\circ{}^t u$.

Now suppose that μ is a cylindrical probability on E. Since $e^{-i\langle \xi,x\rangle}$ is a cylinder function the above integral makes sense and defines $\hat{\mu}$ on E'. Another way of looking at this: we have a coherent system of measures μ_F for each finite-dimensional quotient space F of E. The Fourier transform $\hat{\mu}_F$ resides on

F', which can be identified with a finite-dimensional subspace of E'. The functions $\hat{\mu}_F$ fit together to define a function $\hat{\mu}$ on E'. This function is continuous when restricted to any finite dimensional subspace of E'; also $\hat{\mu}(0) = 1$ and $\hat{\mu}$ is of positive type. Conversely, any such function on E' is the Fourier transform of a cylindrical probability.

Note that $\hat{\mu}(\xi) = (\xi\mu)\hat{\ }(1)$. If we apply this to $\mu = \Gamma$, the Gauss law on H, we find

$$\hat{\Gamma}(\xi) = e^{-\frac{1}{2}|\xi|^2} .$$

Note that, even though $\hat{\Gamma}$ is a *norm*-continuous function, Γ is *not* a Radon measure.

Now consider the space $L^p(\Omega,\mathcal{O},\lambda)$, $0 < p \leq 2$. Define

$$F(f) = e^{-\|f\|^p}$$

or

$$F(f) = \exp\left(-\int_\Omega |f(\omega)|^p \lambda(d\omega)\right) .$$

8.1. <u>Proposition</u>. For $0 < p \leq 2$ the above function is of positive type on $L^p(\Omega,\mathcal{O},\lambda)$.

<u>Proof</u>: We have to show that if $f_1,f_2,\ldots,f_n \in L^p$ and $z_1,z_2,\ldots,z_n \in \mathbb{C}$ then

$$\sum_{ij} \exp(-\|f_i-f_j\|^p_{L^p})z_i\bar{z}_j \geq 0 .$$

It is obviously enough to do this for step functions f_i. Thus we fix our attention on the finite dimensional subspace of L^p defined by a suitable partition $\{\Omega_k\}_1^N$ of Ω. Now if $f = \sum_{i=1}^N \alpha_n 1_{\Omega_n}$ then

$$\|f\|_p^p = \sum_{n=1}^N |\alpha_n|^p \lambda(\Omega_n) .$$

Thus (with $\lambda(\Omega_n) = c_n$) our task is to prove that $F(\alpha_1,\alpha_2,\ldots,\alpha_N) = \exp(-\sum_1^N c_n|\alpha_n|^p)$ is of positive type. But this is just the product of N p-Gauss laws. So we are reduced to the case $N = 1$, for which we are assuming this result. ∎

Remark: Conversely, let E be a Banach space, norm $\|\cdot\|$. In order that E be isometric to a subspace of some L^p space, $1 \leq p \leq 2$, it is necessary and sufficient that the function $e^{-\|\xi\|^p}$ be of positive type on E.

Now let p' be the conjugate index to p $(\frac{1}{p'} + \frac{1}{p} = 1)$. On $L^{p'}$ there exists the analogue Γ_p of p-Gauss law, with $\hat{\Gamma}_p(\xi) = e^{-|\xi|_p^p}$ for $\xi \in L^p$ $(= (L^{p'})')$, $1 < p \leq 2$.

Note: For $p = 1$ we get a cylindrical probability on $\sigma(L^\infty, L^1)$, so that the dual is L^1; this is the analogue of the Cauchy distribution.

For $0 < p < 1$ we can construct Γ_p on the *algebraic* dual $(L^p)^*$ of L^p, with the topology $\sigma((L^p)^*, L^p)$.

More generally, consider a pair of spaces in duality, (E,E'), with $E' \subseteq L^p$. Then we have a cylindrical probability Γ_p on $\sigma(E,E')$, the analogue of the p-Gauss law. Exactly as with the 2-Gauss law, we have

$$\xi(\Gamma_p) = |\xi|\cdot\gamma_p \quad .$$

We now turn to a generalization of the notion of p-summing map. First recall that a *quasi-norm* on a vector space V is a non-negative positively homogeneous function $\|x\|$ such that, for some r, $0 < r \leq 1$, we have

$$\|x+y\|^r \leq \|x\|^r + \|y\|^r \quad .$$

We then say that the space V is r-*normed*.

Note: An r-normed space is s-normed for $0 < s \leq r \leq 1$. Indeed,

$$\|x+y\|^s = (\|x+y\|^r)^{s/r} \leq (\|x\|^r + \|y\|^r)^{s/r}$$

$$\leq (\|x\|^r)^{s/r} + (\|y\|^r)^{s/r} = \|x\|^s + \|y\|^s \quad .$$

A 1-normed space is, of course, just a normed space.

If $0 < r \leq 1$, the space L^r is r-normed, because

$$\int (f+g)^r d\lambda \leq \int |f|^r d\lambda + \int |g|^r d\lambda \quad .$$

Suppose now that (E,E') are in separated duality and that E' is quasi-normed. We also require that the unit ball in E' be weak*-bounded. We can then define $\|e\|_p^* = \sup_{|\xi|\leq 1} \|\langle \xi,e\rangle\|_p$ for a sequence $e = \{e_n\}$ in E.

Now suppose $u\colon E \to F$ where F is quasi-normed. We can then define p-*summing* as before. Then the Pietsch factorization theorem generalizes. In this setting the unit ball of E' need not be compact, but its closure in the algebraic dual E^* is compact, and supports the Pietsch measure.

We can also extend the notion of p-*Radonifying map*. We consider (E,E') in separated duality with a quasi-norm on E'. Thus we have a notion of scalar order for cylindrical probabilities on E. We want $u\colon E \to F$ where F is quasi-normed with separating dual F'. The map u is supposed to be linear and continuous from $\sigma(E,E')$ to $\sigma(F,F')$.

Then, with appropriate modifications, virtually all the previous theorems generalized to this setting.

8.2. Theorem: Suppose that $0 < p \leq 2$ and $E' \subseteq L^p$, so that we have Γ_p on $\sigma(E,E')$. Let $F \subseteq L^q$, where $0 < q \leq 2$, so that Γ_q is defined on $\sigma((L^q)^*,L^q)$. Then the following six properties are equivalent.

(1) $u\colon E \to F$ is completely summing: $\sigma(E,E') \to F$.

(2) For some $r < \bar{p}$, u is r-summing. (Here $\bar{p} = p$ if $p < 2$, while $\bar{2} = \infty$.)

(3) $u(\Gamma_p)$ is Radon on $\sigma(F'',F')$. (Note that Γ_p *is* of *scalar order* r if $r < p$. For $\|\zeta\cdot\Gamma_p\|_r = |\zeta|\|\gamma_p\|_r$, and $\|\gamma_p\|_r < \infty$. So $\|\Gamma_p\|_r^* = \|\gamma_p\|_r$.)

(4) ${}^t u$ is completely summing: $\sigma(F',F) \to \sigma(E',E)$.

(5) For some $r < \bar{q}$, ${}^t u$ is r-summing.

(6) ${}^t u(\Gamma_q)$ is Radon on $\sigma(E^*,E)$.

Notes: (1) The algebraic duals are unnecessary if p and q are 1. If $p = 1$ we may take $E = L^\infty$, $E' = L^1$; E is the dual of E'.

(2) Concerning (3), we amplify what we mean by a Radon measure on $\sigma(F'',F')$. The "unit ball" in F'' is defined to be the closure of the unit ball in F; this may be very small, of course, since balls in F may not be convex. But we require

that a Radon measure on F'' be carried by the union $\bigcup_{n=1}^{\infty} n\bar{B}$ where $\bar{B}$ is the closure in F'' of the unit ball B of F.

Similarly in (6) we require that a Radon measure be carried by $\bigcup_{n=1}^{\infty} n\bar{B}'$, where $\bar{B}'$ is the closure in the algebraic dual E^* of the unit ball B' in E'.

(3) In the general case it is necessary to assume a suitable *approximation property* for (E,E') and (F,F'). It may be dropped for $p = 1$, $E = L^{\infty}$, $E' = L^1$; or for $p > 1$, E a quotient space of $L^{p'}$, E' its dual, a subspace of L^p. The same for F, F', q.

We have the inequality

$$\pi_r({}^t u) \leq \|u(\Gamma_p)\|_r / \|\gamma_p\|_r$$

provided $r < \bar{p}$; cf. Lecture 7. Moreover

$$\frac{\|u(\Gamma_p)\|_r}{\|\gamma_p\|_r} \leq \frac{\pi_r(u)\|\gamma_p\|_r}{\|\gamma_p\|_r} = \pi_r(u) \ .$$

Thus

$$\pi_r({}^t u) \leq \pi_r(u) \ .$$

Suppose in addition that $r < \bar{q}$. Then, by the same reasoning, we get

$$\pi_r(u) \leq \pi_r({}^t u) \ .$$

Accordingly, if $r < \bar{p} \wedge \bar{q}$ (the minimum of $\bar{p}$ and $\bar{q}$), we have

$$\pi_r({}^t u) = \pi_r(u) \ .$$

(This may seem odd: even though u and ${}^t u$ are completely summing, we have equality of their π_r-norms only for $r < \bar{p} \wedge \bar{q}$; we don't know what happens for larger values of r.)

Now suppose that $r \leq s < \bar{p} \wedge \bar{q}$. Then

$$\pi_r({}^t u) \leq \frac{\|u(\Gamma_p)\|_r}{\|\gamma_p\|_r} \leq \frac{\|u(\Gamma_p)\|_s}{\|\gamma_p\|_r} \leq \frac{\pi_s(u)\|\gamma_p\|_s}{\|\gamma_p\|_r} = \pi_s({}^t u)\|\gamma_p\|_s / \|\gamma_p\|_r \ .$$

Thus, if $r \leq s < \bar{p} \wedge \bar{q}$, we have

$$\pi_r(u) \leq \pi_s(u)\|\gamma_p\|_s/\|\gamma_p\|_r \quad .$$

(Of course it is trivial that $\pi_s(u) \leq \pi_r(u)$ if $s \geq r$, with no other restrictions.) Note that the constant blows up as $r \to -1$ because $\|\gamma_p\|_{-1} = 0$.

By reversing the above steps, we get an analogous bound

$$\pi_r(u) \leq \pi_s(u)\|\gamma_q\|_s/\|\gamma_q\|_r \quad .$$

This may sometimes be a better estimate.

We are now in a position to prove the *Pietsch conjecture*. We will give Maurey's argument; as mentioned in Lecture 2, Simone Chevet gave a completely different, independent proof prior to that of Maurey.

8.3. Theorem: Let $u: E \to F$ be p-summing for some $p < 1$. Then u is completely summing, i.e. r-summing for all $r > -1$.

Proof: (1) Suppose $p < 1$ and $\alpha \in \ell^p$. Then the map $(\alpha): \ell^\infty \to \ell^p$ is completely summing, where we use the $\sigma(\ell^\infty,\ell')$ topology on ℓ^∞.

To see this we apply Theorem 8.2. Note first that $(\ell^p)' = \ell^\infty$ because $p < 1$. So the dual separates ℓ^p. All the needed approximation properties are present. Now (α) is p-summing and $p < 1$, so property (2) of Theorem 8.2 holds. Accordingly (α) is completely summing.

Moreover, if $r > -1$, we have (writing $u = (\alpha)$)

$$\pi_r(u) \leq \|\alpha\|_{\ell^p}\|\gamma_1\|_p/\|\gamma_1\|_r \quad .$$

(2) Next consider $(\alpha): L^\infty \to L^p$ for a general measure space. L^∞ is separated by L^1, so the $\sigma(L^\infty,L^1)$ topology is Hausdorff. However, in general L^p is *not* separated by its dual. But this is not a problem if the measure is *discrete*; we then reduce to case (1).

The discrete case is actually enough to take care of the general case. For in proving that a map is completely summing it is enough to have uniform estimates for all finite dimensional subspaces of L^∞. This is the case here since the constants in (1) are universal. Accordingly (α) is completely summing.

(3) The general case: suppose that $u: E \to F$ is p-summing for some $p < 1$. Then we have a Pietsch factorization of u:

$$\begin{array}{ccccc} E & \longrightarrow & L^{\infty} & \xrightarrow{(\alpha)} & L^{p} \\ & \searrow & & \nearrow & \\ & & S & \longrightarrow & F \end{array}$$

Since (α) is completely summing, u is completely summing.

Moreover, for $r < p < 1$ we have the estimate

$$\pi_r(u) \leq \|\alpha\|_{L^p} \|\gamma_1\|_p / \|\gamma_1\|_r \; . \qquad \blacksquare$$

Lecture 9. p-Pietsch Spaces. Application: Brownian Motion

In Theorem 8.2, suppose that we keep the hypothesis on E (so that $E' \subseteq L^p$, $p \leq 2$), but drop the hypothesis on F. Then it is still the case that (1) ⇒ (2) ⇒ (3) ⇒ (4) ⇒ (5). However, lacking Γ_q, we have no way to go from (5) to (6). None of the reverse implications are valid except that (2) ⇒ (1).

Reason: We may suppose that $1 \leq r < \bar{p}$, since if u is r-summing for an $r < 1$ then the Pietsch conjecture (Theorem 8.3) implies that u is completely summing. Then we have a factorization of u:

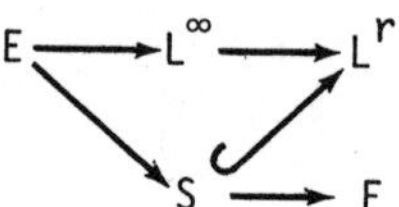

Consider the map $E \to L^{\infty} \to L^{r}$. We apply Theorem 8.2 to this map. Since it is r-summing for $r < \bar{p}$, it is completely summing. Accordingly u is completely summing, with $\pi_s(u) \leq \pi_r(u)\|\gamma_p\|_r/\|\gamma_p\|_s$ for $s \leq r$.

The above reasoning applies to all the spaces $E = (L^r)'$, with L^r as dual. The same for E a quotient of $(L^r)'$, for then E' is a subspace of L^r. This motivates the following definition.

Definition: E is p-*Pietsch* provided that every p-summing map from E into a Banach space is completely summing.

For example, every Banach space E is $(1-\varepsilon)$-Pietsch, by virtue of the Pietsch conjecture.

By a "closed graph" argument, if E is p-Pietsch it follows that there are constants $C_{r,p}$ $(r \leq p)$ such that $\pi_r(u) \leq C_{r,p}\pi_p(u)$ for all $u: E \to X$, X Banach.

Suppose that $1 \leq p \leq 2$. Then our reasoning given above shows that $L^{p'}$ is $(p-\varepsilon)$-Pietsch. Moreover the constant $C_{r,s}$ is $\leq \|\gamma_p\|_s/\|\gamma_p\|_r$. Likewise, every quotient of $(L^p)'$ is $(p-\varepsilon)$-Pietsch.

L^2 is p-Pietsch for $p \leq 2$, so L^2 is 2-Pietsch.

9.1. Proposition: If E is k-Pietsch then so is every subspace of E.

Proof: Suppose $E_0 \subseteq E$. Consider $u: E_0 \to F$, a k-summing map. We have a factorization:

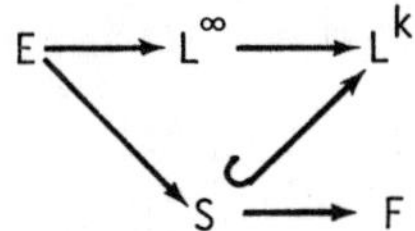

So it suffices to prove that the map $E_0 \to L^\infty \to L^k$ is completely summing. Now, by a theorem of Nachbin, L^∞ is an "injective" Banach space. That is, it has the "Hahn-Banach property", so that the map $u: E_0 \to L^\infty$ extends to a map $E \to L^\infty$ with the same norm. Thus we have a k-summing map $E \to L^\infty \to L^k$, and this is completely summing because E is k-Pietsch. So the restriction to E_0 is completely summing. ■

As a corollary, every subspace of a quotient of $L^{p'}$ is $(p-\varepsilon)$-Pietsch, $1 \leq p \leq 2$.

Note: For $p > 2$, $L^{p'}$ is 2-Pietsch. But we will not be able to prove this until we have developed the notions of "type" and "cotype".

Applications: Sobolev Spaces and Brownian Motion

Over $\mathbb{R}^n$ or $\mathbb{T}^n$, consider the space $(L^a)^\alpha$ of distribution whose derivatives up to order α lie in L^a. (Fractional derivatives can be defined in terms of Fourier series.) Thus $T \in (L^a)^\alpha$ means $D^\alpha T \in L^a$.

9.2. Sobolev Theorem: Suppose that α, β are real and $1 \leq a,b \leq +\infty$. Regard L^∞ as C. Then (at least locally on $\mathbb{R}^n$, globally on $\mathbb{T}^n$), if $\frac{\alpha-\beta}{n} > (\frac{1}{a}-\frac{1}{b})_+$ then $(L^a)^\alpha \subseteq (L^b)^\beta$.

Here $t_+ = \max(t,0)$, the "positive part" of t. For example, if $\frac{s}{n} > \frac{1}{2}$ then $(L^2)^s \subseteq C$; here the condition is necessary as well as sufficient.

It is natural, from our point of view, to ask when the injection $(L^a)^\alpha \to (L^b)^\beta$ is p-summing. One would like necessary and sufficient conditions on the parameters (n,a,α,b,β,p). The following is a sufficient condition.

9.3. Proposition: Suppose that $\frac{\alpha-\beta}{n} > \frac{1}{a} + (\frac{1}{p}-\frac{1}{b})_+$. Here we take $p \geq 1$. Then the injection $(L^a)^\alpha \to (L^b)^\beta$ is summing.

Proof: We have

$$(L^a)^\alpha \xrightarrow{D^\gamma} (L^a)^{\alpha-\gamma} \hookrightarrow L^\infty ,$$

where the final inclusion is valid by Sobolev's Theorem provided that $\frac{\alpha-\gamma}{n} > \frac{1}{a}$. We have the natural inclusion $L^\infty \hookrightarrow L^p$. Finally we have

$$L^p \xrightarrow{D^{-\gamma}} (L^p)^\gamma \quad .$$

We wish to end up with $(L^p)^\gamma \subseteq (L^b)^\beta$. This will be the case if $\frac{\gamma-\beta}{n} > (\frac{1}{p}-\frac{1}{b})_+$. Thus we need the parameter γ to satisfy the two inequalities

$$\frac{\alpha-\gamma}{n} > \frac{1}{a}$$

and

$$\frac{\gamma-\beta}{n} > (\frac{1}{p}-\frac{1}{b})_+ \quad .$$

Such a γ exists provided $\frac{\alpha-\beta}{n} > \frac{1}{a} + (\frac{1}{p}-\frac{1}{b})_+$. ■

The method used in the above proof is very crude. Using Theorem 8.2 one can get a more refined result.

9.4. Theorem: If $\frac{\alpha-\beta}{n} > \frac{1}{b'} \vee (\frac{1}{2} \wedge \frac{1}{a})$ then the injection $(L^a)^\alpha \to (L^b)^\beta$ is completely summing.

This problem has been studied in detail by Beauzamy. He divides the parameters into 10 cases, and gets necessary and sufficient conditions in all but one case, where a gap remains.

Now we turn to our application to stochastic processes. "White noise" is given by Γ_2 on L^2. "Brownian motion" is $\int_0^t f(s)ds$ if f is white noise. So Brownian motion is described by Γ_2 on $(L^2)^1$. (For more details on stochastic processes see the next lecture.)

We shall prove that Brownian motion is a.s. continuous and can be realized by a Radon measure on C. To see this we just have to show that the injection $(L^2)^1 \hookrightarrow C$ radonifies the Gauss measure Γ_2. It is enough to prove that this map is p-summing for some p.

Actually we will do more; we will prove that Brownian motion is a.s. β-Hölder for $\beta < \frac{1}{2}$, a famous result of N. Wiener. So consider the inclusion $(L^2)^1 \hookrightarrow C$. (It is the same if C^β is replaced by $(L^\infty)^\beta$ since we are considering *all* $\beta < \frac{1}{2}$.) For the inclusion to be p-summing, we need (by Proposition 9.3)

$$1 - \beta > \frac{1}{2} + \frac{1}{p}$$

or

$$\beta < \frac{1}{2} - \frac{1}{p} .$$

Since p can be taken as large as we wish, we only need $\beta < \frac{1}{2}$.

Note: The assertion about Brownian motion is definitely false for $\beta = \frac{1}{2}$, and this shows that the sufficient condition in Proposition 9.3 can't be improved too much.

One can use the same methods to deal with the analogues of Brownian motion defined by the p-Gauss law.

Lecture 10. More on Cylindrical Measures and Stochastic Processes

A stochastic process may be described as follows. Let T be an index set. Typically $T \subseteq \mathbb{R}$, and the elements in T are "times". To each $t \in T$ there corresponds a random variable X_t. The process is specified by giving the joint distributions of $(X_{t_1}, X_{t_2}, \ldots, X_{t_n})$ over $(\Omega, \mathcal{O}, \lambda)$. Usually one is only given the joint laws on $\mathbb{R}^n$; one has to construct the space Ω. Kolmogoroff's approach leads to $\Omega = \mathbb{R}^T$ with a certain $\mathcal{O}$, λ.

Another method goes as follows. Let $\bar{\mathbb{R}} = [-\infty, \infty]$, a compactification of $\mathbb{R}$. For $\{t_1, t_2, \ldots, t_n\} \subset T$ we have a probability measure on $\bar{\mathbb{R}}^{t_1} \times \cdots \times \bar{\mathbb{R}}^{t_n} = \mathbb{R}^{\{t_1, \ldots, t_n\}}$, and the coherence condition is satisfied, so that we can take the projective limit to get a Radon probability measure on $\bar{\mathbb{R}}^T$.

We can also embed the index set T in a larger index set with a vector space structure. Thus let $\mathbb{R}^{(T)}$ be the direct sum of copies of $\mathbb{R}$ indexed by T. A typical member of $\mathbb{R}^{(T)}$ is a formal linear combination $c_1 t_1 + \cdots + c_n t_n$, $c_i \in \mathbb{R}$, $t_i \in T$. If $f(t)$ is a stochastic process indexed by T, we can define $f(c_1 t_1 + \cdots + c_n t_n) = c_1 f(t_1) + \cdots + c_n f(t_n)$, thus defining a new process with index set $\mathbb{R}^{(T)}$. Thus we have a linear map $\mathbb{R}^{(T)} \to L^0(\Omega, \mathcal{O}, \lambda)$. (Here L^0 is the space of measurable functions, equipped with the topology of convergence in measure.) In this way the case of non-linear processes reduces to the linear case, over a huge vector space.

Linear processes: Suppose that E is a Banach space and that we have a linear map

$$f\colon E' \to L^0(\Omega, \mathcal{O}, \lambda) \quad .$$

This defines a *cylindrical probability* μ on E. Construction: For $\xi_1, \xi_2, \ldots, \xi_n \to E'$ we have the linear map $(\xi_1, \ldots, \xi_n)\colon E \to \mathbb{R}^n$, to which we must associate a measure $(\xi_1, \ldots, \xi_n)(\mu)$ on $\mathbb{R}^n$. We simply define

$$\begin{aligned}(\xi_1, \ldots, \xi_n)(\mu) &= (f(\xi_1), \ldots, f(\xi_n))(\lambda) \\ &= \text{the joint law of } f(\xi_1), \ldots, f(\xi_n) \quad .\end{aligned}$$

The coherence condition is satisfied because f is linear.

Let us calculate the scalar order of μ. We have

$$\|\mu\|_p^* = \sup_{|\xi|\leq 1} \|\xi(\mu)\|_p$$

$$= \sup_{|\xi|\leq 1} \|f(\xi)(\lambda)\|_p$$

$$= \sup_{|\xi|\leq 1} \|f(\xi)\|_{L^p(\Omega,\mathcal{O},\lambda)}$$

so that $\|\mu\|_p^*$ is just the norm of f as a map from E' into $L^p(\Omega,\mathcal{O},\lambda)$:

$$\|\mu\|_p^* = \|f\|_{\mathcal{L}(E';L^p(\Omega,\mathcal{O},\lambda))} .$$

Suppose now that $u: E \to F$ is a continuous linear map. Then we have

$$F' \xrightarrow{\;{}^t u\;} E' \xrightarrow{\;f\;} L^0(\Omega,\mathcal{O},\lambda) ,$$

defining the composite process $f\circ{}^t u$ over F'. The corresponding cylindrical measure on F is of course $u\cdot\mu$.

The process $f': E' \to L^0(\Omega',\mathcal{O}',\lambda')$ is said to be *equivalent* to f provided it has the same joint laws, i.e. leads to the same cylindrical measure on E.

Next, suppose that we are given a space E with a cylindrical probability μ. There is a corresponding stochastic process f. To see this, take Ω to be $\bar{\mathbb{R}}^{E'}$. Then for $\xi_1,\ldots,\xi_n \in E'$ consider $\bar{\mathbb{R}}^{\{\xi_1,\xi_2,\ldots,\xi_n\}}$ with measure $\mu_{(\xi_1,\ldots,\xi_n)}$; these measures form a projective system for the compact set Ω. The coherence condition is automatic here, since we started with a cylindrical probability μ. Thus we have a Radon probability λ on Ω with the above joint distributions.

To finish the construction we must give a linear map $f: E' \to L^0(\Omega,\mathcal{O},\lambda)$. If $\xi \in E'$ we define $f(\xi): \Omega \to \bar{\mathbb{R}}$ to be the projection π_ξ on the ξ-coordinate factor of Ω. To show that f is linear, we have to prove that $\pi_{\xi+\eta} = \pi_\xi + \pi_\eta$. Of course, this is false! However, it is true that $\pi_{\xi+\eta} = \pi_\xi + \pi_\eta$ λ-a.e. This follows from the coherence condition on μ. Indeed, consider the map $(\xi+\eta,\xi,\eta)$: $E \to \mathbb{R}^3$. This leads to the measure $\mu_{(\xi+\eta,\xi,\eta)}$ on $\mathbb{R}^3$. We claim that this measure is carried by the plane $x_1-x_2-x_3 = 0$; this will prove that $\pi_{\xi+\eta} = \pi_\xi + \pi_\eta$

a.e. To verify the claim, define $T(x_1,x_2,x_3) = x_1-x_2-x_3$. Then the diagram

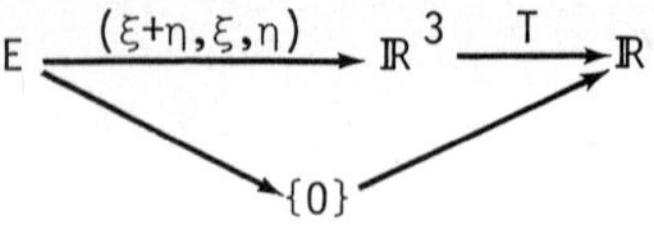

commutes. Therefore $T(\mu_{(\xi+\eta,\xi,\eta)}) = \delta_{\{0\}}$, which proves that $\mu_{(\xi+\eta,\xi,\eta)}$ is supported on the plane $T^{-1}(0)$.

<u>Example</u>: Let H be a Hilbert space, and consider the Gauss cylindrical probability Γ_2 on H. We can realize Γ_2 by a stochastic process

$$f\colon H' \to L^0(\Omega,\mathcal{O},\lambda) \quad .$$

The map f has some rather exceptional properties. First of all, for $\xi \in H'$, $f(\xi)$ is a Gaussian random variable with parameter $|\xi|$, i.e. $f(\xi)(\lambda) = |\xi|\cdot\gamma_2$. As a consequence,

$$\|f(\xi)\|_{L^p(\Omega,\mathcal{O},\lambda)} = |\xi|\cdot\|\gamma_2\|_p \quad .$$

Because $\|\gamma_2\|_2 = 1$, the map f is an isometry from H' into $L^2(\Omega,\mathcal{O},\lambda)$. More generally, $\frac{1}{\|\gamma_2\|_p} f$ is an isometric embedding of the Hilbert space H' into $L^p(\Omega,\mathcal{O},\lambda)$. Denote the range of f by $K \subseteq L^2(\Omega,\mathcal{O},\lambda)$. On this closed subspace of L^2, *all* the L^p topologies are the same. (This includes the L^0 topology, because, for Gaussian random variables, convergence in law implies L^2 convergence.) Moreover all the L^p norms are proportional on K.

Since H is separable, Ω is separable. Hence K can be realized as a closed subspace of $L^2(0,1)$ with the above properties. (It is not at all easy to see directly that a non-trivial subspace with these properties exists.)

We can of course repeat the same construction starting with Γ_p, the p-Gauss law on $\sigma((L^p)^*,L^p)$, $0 < p \leq 2$. We thus have a linear map $f\colon L^p \to L^0(\Omega,\mathcal{O},\lambda)$ which realizes Γ_p. Thus, for $\xi \in L^p$, we have

$$f(\xi)(\lambda) = |\xi|\cdot\gamma_p = \xi(\Gamma_p) \quad .$$

Therefore, if $q < p$, we have

$$\|f(\xi)\|_{L^q(\Omega,\mathcal{O},\lambda)} = |\xi| \cdot \|\gamma_p\|_q \quad .$$

Conclusions: If $q < p \leq 2$, then L^p can be isometrically embedded in L^q. Also, on the range of f, a subspace of L^0, all L^q norms are proportional for $q < p$. Thus L^p sits in L^q as a closed subspace $(q < p)$ with all norms proportional.

One might ask whether L^p (for p finite) can be replaced by L^∞. Does there exist in L^∞ a non-trivial subspace E for which the L^∞ topology coincides with, e.g., the L^2 topology? The answer is no. For let $E \subseteq L^\infty$ be a closed subspace whose topology is the same as the L^p topology for some $p < \infty$. By Nachbin's Theorem, E is a direct summand of L^∞. Thus we have $E \overset{i}{\to} L^\infty \overset{\pi}{\to} E = L^1$. (N.B. The topology on E is the same as the L^1 topology by our hypothesis.) But we shall eventually see that every continuous linear map $\pi: L^\infty \to L^1$ is 2-summing. Hence $\pi \circ i$ = identity on E is 2-summing, so that E is finite dimensional.

III. Types and Cotypes

Let ε_n, $1 \leq n < +\infty$, be Rademacher random variables (independent with $\varepsilon_n = \pm 1$ with probability 1/2). For (x_n) a sequence of real numbers, we have a dichotomy: either

$$\sum_{n=1}^{\infty} |x_n|^2 < +\infty$$

and

$$(*) \qquad \sum_{n=1}^{\infty} \varepsilon_n x_n$$

is a.s. convergent, or $\sum_1^{\infty} |x_n|^2 = +\infty$ and $(*)$ is a.s. divergent. As indicated in Lecture 1, we want to generalize to sequences (x_n) in a Banach space E. When is it true that $\sum_1^{\infty} |x_n|^2 < +\infty \Leftrightarrow \sum_1^{\infty} \varepsilon_n x_n$ is a.s. convergent? A theorem of Kwapien tells us that this is so if and only if E can be renormed to be a Hilbert space.

Definition: E is of *type* p provided that

$$\left(\sum_1^{\infty} |x_n|^p\right)^{1/p} < +\infty \Rightarrow \sum_1^{\infty} \varepsilon_n x_n \text{ is a.s. convergent.}$$

(The line $\mathbb{R}$ is of type 2.)

Remarks: Every Banach space is of type 1, trivially. So we are interested only in $p > 1$. Also, we must take p to be ≤ 2. For if $E \neq \{0\}$ we can take all the x_n on the same straight line, and for $p > 2$ the line is not of type p.

It is obvious that type p implies type q if $q < p$. So there is an *interval of types* $[1,p_0|$ for E, where p_0 may or may not belong to the interval.

The notion of *cotype* is defined in terms of the reverse implication $\Leftarrow$ (cf. Lecture 1). We must consider only cotypes $q \geq 2$, and every Banach space is of cotype $+\infty$.

We can also define the *type of a linear map* $u: E \to F$.

Definition: A continuous linear map is of type p $(0 < p \leq 2)$ provided that

$$\left(\sum_1 |x_n|^p\right)^{1/p} < +\infty \Rightarrow \sum_1^{\infty} u(x_n)\varepsilon_n \text{ is a.s. convergent.}$$

If one factor in a product of continuous linear maps is type p, so is the whole product.

A space E is of type p if and only if the identity on E is a map of type p.

Properties: (1) L^p, $1 \leq p \leq 2$, is of type p, and no better (unless finite dimensional).

(2) L^r, $2 \leq r < +\infty$, is of type 2.

These properties will be proved in Lecture 11.

(3) L^∞ is type 1, and no better, just like L^1. This can be understood as follows. "Type" is inherited by subspaces (trivially), and also by quotient spaces. (Easy proof: A sequence $\dot{x}_n$ in a quotient space can be lifted to x_n in E of nearly the same norm, so that the corresponding series have the same convergence properties.) This is why L^∞ and L^1 are as bad as possible; for "everything" is a subspace of L^∞ and a quotient of L^1.

(4) If E' is of type p, then E is (p-ε)-Pietsch, except for p = 2, when we get that E is 2-Pietsch.

Application: If $1 \leq r \leq 2$ then $L^{r'}$ is (r-ε)-Pietsch,

For if $E = L^{r'}$ then $E' = L^r$ is of type r, so E is (r-ε)-Pietsch.

This was proved earlier in Lecture 9 using Theorem 8.2 ("six equivalent properties").

Lecture 11. Kahane Inequality. The Case of L^p. Z-type

The following key inequality is due to *Kahane*.

11.1. Theorem: Let (ε_n) be Rademacher variables and (y_n) a sequence in a Banach space. If the series $\sum_n \varepsilon_n y_n$ converges a.s. then, for all $r < +\infty$,

$$\left(E_\varepsilon |\sum_n \varepsilon_n y_n|^r\right)^{1/r} < +\infty \quad .$$

Moreover there are universal constants $C_{s,r}$ such that

$$\left(E_\varepsilon|\sum_n \varepsilon_n y_n|^s\right)^{1/s} \le C_{s,r}\left(E_\varepsilon|\sum_n \varepsilon_n y_n|^r\right)^{1/r} .$$

(Of course this is of interest only for $s > r$.)

We omit the proof (see references).

11.2. Corollary: A linear map $u: E \to F$ is type p if and only if, for some (hence every) finite q, there exists a constant $\tau_{p,q}$ such that, for all finite sequences (x_n),

$$\left(E_\varepsilon|\sum\varepsilon_n u(x_n)|^q\right)^{1/q} \le \tau_{p,q}\left(\sum_n |x_n|^p\right)^{1/p} .$$

(Thus we need not consider a.s. convergence in discussing the type of a linear map.)

11.3. Theorem: If $1 \le p \le 2$, then L^p is of type p.

Proof: Denote the underlying measure by dt. Consider a finite sequence of functions (x_n) in L^p. We will apply Corollary 11.2 with $q = p$. So we form the sum

$$|\sum_n \varepsilon_n x_n(t)|^p_{L^p} = \int |\sum_n \varepsilon_n x_n(t)|^p dt .$$

Thus

$$E_\varepsilon|\sum_n \varepsilon_n x_n|^p = E_\varepsilon \int |\sum_n \varepsilon_n x_n(t)|^p dt = \int dt\ E_\varepsilon|\sum_n \varepsilon_n x_n(t)|^p$$

by Fubini's Theorem. Now, with t fixed, we use the fact that $\mathbb{R}$ is type 2, and hence type p, together with Kahane's inequality, to write

$$E_\varepsilon|\sum_n \varepsilon_n x_n(t)|^p \le C \sum_n |x_n(t)|^p$$

where C is a universal constant. Integrating with respect to t, we get

$$\left(E_\varepsilon|\sum_n \varepsilon_n x_n|^p\right)^{1/p} \le C^{1/p}\left(\sum_n |x_n|^p\right)^{1/p} .$$

Thus L^p is of type p. ∎

11.4. Theorem: If an L^p space is infinite dimensional, it is not of type $q > p$.

Proof: In this case, L^p contains a subspace isometric to ℓ_p, so it suffices to show that the latter space is not of type $q > p$.

Let (η_n) be the standard unit vector basis for ℓ_p and consider $x_n = \alpha_n \eta_n$, $\alpha_n \in \mathbb{R}$. Suppose that $\sum_n \varepsilon_n \alpha_n \eta_n$ is a.s. convergent. Note that this series converges for one choice of signs ε_n $\Leftrightarrow$ it converges for *all* choices of signs $\Leftrightarrow$ $\sum_n |\alpha_n|^p < +\infty$. The latter is certainly not implied by the condition $\sum_n |\alpha_n|^q < +\infty$ if $q > p$. ∎

11.5. Theorem: If $2 \leq r < +\infty$, then L^r is of type 2.

Remark: It follows from this and Theorem 11.3 that in general L^r is of type $r \wedge 2$, except for L^∞, which is of type 1.

Proof: Consider a finite system of functions (x_n) in L^r. We will apply Corollary 11.2 with $p = 2$ and $q = r$. Now

$$\left(E_\varepsilon \int |\sum_n \varepsilon_n x_n(t)|^r dt\right)^{1/r} = \left(\int dt\ E_\varepsilon |\sum_n \varepsilon_n x_n(t)|^r\right)^{1/r} .$$

Since $\mathbb{R}$ is of type 2, Kahane's inequality implies that

$$E_\varepsilon |\sum_n \varepsilon_n x_n(t)|^r \leq C\left(\sum_n |x_n(t)|^2\right)^{r/2} .$$

So the right side above is

$$\leq C\left(\int dt \left(\sum_n |x_n(t)|^2\right)^{r/2}\right)^{1/r}.$$

The latter is a "$L_t^r(\ell^2)$" norm.

Now, by Minkowski's inequality, we can compare $L_x^a(L_y^b)$ with $L_y^b(L_x^a)$.

Consider the $L_x^a(L_y^b)$ norm of $f(x,y)$: This is

$$\left[\int \left(\int |f(x,y)|^b dy\right)^{a/b} dx\right]^{1/a} .$$

Now, if $a \geq b$, the x integral is the $1/b$ power of the $L_{a/b}$ norm of a linear combination of functions $|f(\cdot,y)|^b$, whence the above expression is

$$\leq \left[\int dy \left(\int |f(x,y)|^a dx\right)^{b/a}\right]^{1/b} ,$$

which is the $L_y^b(L_x^a)$ norm of f.

Applying this with $a = r$ and $b = 2$, we find

$$\left(\int dt\left(\sum_n |x_n(t)|^2\right)^{r/2}\right)^{1/r} \le \left(\sum_n\left(\int dt|x_n(t)|^r\right)^{2/r}\right)^{1/2}$$

$$= \left(\sum_n |x_n|^2_{L^r}\right)^{1/2} .$$

Thus

$$\left(E_\varepsilon|\sum\varepsilon_n x_n|^r_{L^r}\right)^{1/r} \le C\left(\sum_n |x_n|^2_{L^r}\right)^{1/2} ,$$

which shows that L^r is of type 2. ∎

In our discussion of "type" the Rademacher random variables played a special role. It is of interest to investigate what happens when they are replaced by some other system of independent R.V.'s. Thus, let (Z_n) be a sequence of symmetric independent R.V.'s (not necessarily identically distributed). We say that a space E is *type* p *relative to* (Z_n) provided that, for every sequence (x_n) in E,

$$\left(\sum_n |x_n|^p\right)^{1/p} < +\infty \Rightarrow \sum_n x_n Z_n \text{ converges in mean of order p.}$$

Note: We specify L^p mean convergence rather than a.s. convergence since we lack a Kahane inequality in this setting. Of course, since the Z_n are independent, mean convergence implies a.s. convergence.

11.6. Theorem: Assume that the Z_n satisfy

$$E(|Z_n|^p)^{1/p} \le a < +\infty$$

where a is independent of n. Then if the space E is ε-type p, E is also Z-type p.

Proof: On $\Omega' \times \Omega$ consider the system of independent R.V.'s $\varepsilon_n(\omega')Z_n(\omega)$. Because Z_n is symmetric, this $\varepsilon_n Z_n$ system has the same distribution law as Z_n. Accordingly, for $x_n \in E$,

$$\left(E_\omega\|\sum_n(\omega)x_n\|^p_E\right)^{1/p} = \left(E_{\omega',\omega}\|\sum_n\varepsilon_n(\omega')Z_n(\omega)x_n\|^p_E\right)^{1/p} .$$

Because E is assumed to be of ε-type p, the latter sum is

$$\leq \tau_{p,p}\left(E_\omega \sum_n |Z_n(\omega)|^p \|x_n\|_E^p\right)^{1/p}$$

where $\tau_{p,p}$ is the "Rademacher" constant of Corollary 11.2. Since $E_\omega|Z_n(\omega)|^p \leq a^p$, this in turn is bounded by

$$\tau_{pp} a\left(\sum_n \|x_n\|_E^p\right)^{1/p} .$$

Thus E is Z-type p. ∎

Conversely, we will give conditions which guarantee that Z-type p implies ε-type p.

Lecture 12. Kahane Contraction Principle. p-Gauss Type The Gauss Type Interval is Open

To develop conditions that insure that Z-type p implies ε-type p, we will need the following "contraction principle" due to *Kahane*.

12.1. Theorem (Kahane): Let $U_1, U_2, \ldots, U_n$ be symmetric independent R.V.'s with values in a Banach space. Let $\alpha_1, \alpha_2, \ldots, \alpha_n$ be real scalars with $|\alpha_i| \leq 1$. Let $p \geq 1$. Then

$$E(|\alpha_1 U_1 + \cdots + \alpha_n U_n|^p) \leq E(|U_1 + \cdots + U_n|^p) .$$

Proof: First take $n = 2$, and suppose that u, v are fixed vectors while α is real, $|\alpha| \leq 1$. Then we have the elementary inequality

$$(*) \qquad \frac{1}{2}(|u+\alpha v|^p + |u-\alpha v|^p) \leq \frac{1}{2}(|u+v|^p + |u-v|^p) .$$

Indeed, the left side is an even, convex function of α, and so is increasing for $0 \leq \alpha \leq 1$.

For general n, introduce Rademacher R.V.'s. We have, for $U_1, \ldots, U_n$ fixed vectors, and $\alpha_1, \ldots, \alpha_n$ real with $|\alpha_j| \leq 1$, the inequality

$$E_\varepsilon|\varepsilon_1\alpha_1U_1+\cdots+\varepsilon_{n-1}\alpha_{n-1}U_{n-1}+\varepsilon_n\alpha_nU_n|^p \le E_\varepsilon|\varepsilon_1\alpha_1U_1+\cdots+\varepsilon_{n-1}\alpha_{n-1}U_{n-1}+\varepsilon_nU_n|^p .$$

This follows from (*) if we first average over ε_n. By iterating this argument, we see that

$$E_\varepsilon|\varepsilon_1\alpha_1U_1+\cdots+\varepsilon_n\alpha_nU_n|^p \le E_\varepsilon|\varepsilon_1U_1+\cdots\varepsilon_nU_n|^p .$$

Next, we observe that $|\alpha_1U_1(\omega)+\cdots+\alpha_nU_n(\omega)|^p$ has the same distribution as $|\alpha_1\varepsilon_1(\omega')U_1(\omega)+\cdots+\alpha_n\varepsilon_n(\omega')U_n(\omega)|$. Hence

$$\begin{aligned} E_\omega|\alpha_1U_1+\cdots+\alpha_nU_n|^p &= E_{\omega',\omega}|\alpha_1\varepsilon_1U_1+\cdots+\alpha_n\varepsilon_nU_n|^p \\ &\le E_{\omega',\omega}|\varepsilon_1U_1+\cdots+\varepsilon_nU_n|^p \\ &= E_\omega|U_1+\cdots+U_n|^p . \quad ⌘ \end{aligned}$$

12.2. <u>Corollary</u>: Under the same hypotheses, suppose that $0 \le |\alpha_n| \le \beta_n$. Then

$$E|\alpha_1U_1+\cdots+\alpha_nU_n|^p + E|\beta_1U_1+\cdots+\beta_nU_n|^p. \quad ⌘$$

We showed in Lecture 11 that if Z_n is a sequence of symmetric, independent R.V.'s with $(E|Z_n|^p)^{1/p} \le a$, then a space of ε-type p is of Z-type p. We have the following result in the reverse direction.

12.3. <u>Theorem</u>: Suppose that the symmetric, independent R.V.'s (Z_n) satisfy the inequality

$$E|Z_n| \ge a > 0$$

for some constant a. Then for $p \ge 1$, every space of Z-type p is also of ε-type p.

(For example, the Z_n can be identically distributed Gaussian R.V.'s.)

<u>Proof</u>. It obviously suffices to establish the inequality

$$(*) \qquad E|\sum_n Z_nx_n|^p \ge a^pE|\sum_n \varepsilon_nx_n|^p .$$

Once again we use the fact that $\varepsilon_n(\omega')|Z_n(\omega)|$ has the same distribution as Z_n. Now

$$E_{\omega',\omega}\left|\sum_n \varepsilon_n(\omega')|Z_n(\omega)|x_n\right|^p = E_{\omega'}\left(E_\omega\left|\sum_n \varepsilon_n|Z_n|x_n\right|^p\right)$$

$$\geq E_{\omega'}\left|E_\omega\left(\sum_n \varepsilon_n(\omega')|Z_n(\omega)|x_n\right)\right|^p$$

$$= E_{\omega'}\left|\sum_n \varepsilon_n(\omega')|Z_n|_{L^1}x_n\right|^p \quad .$$

Now we apply the Kahane contraction principle, noting that $|Z_n|_{L^1} = E|Z_n| \geq a$. Thus the last term above is

$$\geq E_{\omega'}|\sum_n a\varepsilon_n(\omega')x_n|^p$$

$$= a^p E_{\omega'}|\sum_n \varepsilon_n(\omega')x_n|^p \quad .$$

This establishes the inequality. ■

For the case of p-*Gaussian* random variables there is a different notion of type which depends on a "Kahane-type" inequality.

<u>Definition</u>: Let (Z_n) be independent p-Gaussian random variables. The space E is of p-*Gauss type* provided that

$$\left(\sum_n |x_n|^p\right)^{1/p} < +\infty \Rightarrow \sum_n Z_n x_n \text{ converges a.s.}$$

Note that if p is changed in this definition then the underlying sequence (Z_n) of random variables is changed. So there is no obvious way to compare different values of p --but see below.

The following deep result of Hoffman-Jørgensen plays the role of the Shepp-Landau-Fernique Theorem for 2-Gaussian R.V.'s or Kahane's Theorem (11.1) for Rademacher R.V.'s.

12.4. <u>Theorem</u> (Hoffman-Jørgensen): Let (Z_n) be p-Gauss random variables. If the sequence $\sum_n Z_n x_n$ converges a.s. then it converges in means of order q for all $q < \bar{p}$. Hence there exist universal constants $C_{r,q}$ such that, whenever $r,q < p$,

$$\left(E|\sum_n Z_n x_n|^r\right)^{1/r} \leq C_{r,q}\left(E|\sum_n Z_n x_n|^q\right)^{1/q} \quad .$$

Hence, as a corollary, the space E is of type p-Gauss if and only if for any (hence every) $q < p$ there is a constant $\tau_{q,p} = \tau_{q,p}(E)$ such that

$$\left(E\left|\sum_n Z_n x_n\right|^q\right)^{1/q} \leq \tau_{q,p}(E)\left(\sum_n |x_n|^p\right)^{1/p}$$

where Z_n = p-Gauss random variable. (This is analogous to Corollary 11.2 for Rademacher R.V.'s.)

For $p = 2$, we take Z_n = 2-Gauss R.V. Then the two possible notions of type coincide: E is of type 2 relative to (Z_n) if and only if E is of 2-Gauss type. Moreover, by virtue of Theorems 11.6 and 12.3, E is of type 2-Gauss if and only if E is of type 2-Rademacher.

More generally, for arbitrary μ, we have the following result.

12.5. Proposition: If E is of type p-Gauss it is of type p-Rademacher. (The converse is false because the p-Gauss variables do not have finite p-moments, if $p \neq 2$.)

Proof: Suppose E is type p-Gauss. Then there are constants $\tau_{qp}(E)$ such that, if $q < p$.

$$\left(E\left|\sum_n Z_n x_n\right|^q\right)^{1/q} \leq \tau_{qp}(E)\left(\sum_n |x_n|^p\right)^{1/p} .$$

By virtue of the inequality (*) in the proof of Theorem 12.3, the left side is

$$\geq \text{Constant}\cdot\left(E\left|\sum_n \varepsilon_n x_n\right|^q\right)^{1/q} . \quad ■$$

12.6. Corollary: If E is of type p-Gauss it is of type q-Gauss for any $q < p$.

Proof: By Proposition 12.5, p-Gauss ⇒ p-Rademacher. By Theorem 11.3, p-Rademacher ⇒ $(p-\varepsilon)$-Gauss for any $\varepsilon > 0$. ■

Thus the different p-Gauss types *are* comparable after all. This is a result of Maurey and Pisier. They also established the following powerful theorem.

12.7. Theorem (Maurey-Pisier): Let E be a Banach space. Consider the associated type interval $(0,p_0|$, $0 < p_0 \leq 2$. The *Gauss* type interval is *open*, of the form $(0,p_0)$, unless $p_0 = 2$, when either possibility can occur.

That is, if E is of type p-Gauss and $p < 2$, then E is of type $(p+\varepsilon)$-Gauss for some $\varepsilon > 0$. Because p-Gauss $\Rightarrow$ p-Rademacher, it follows that if $p < 2$ then E is of type p-Gauss $\Leftrightarrow$ E is of type $(p+\varepsilon)$-Rademacher for some $\varepsilon > 0$.

(In a sense this eliminates the separate study of Gauss type, although the notion remains useful in the proofs of other theorems.)

There is also a p-Gauss type notion for *linear maps*. In this setting the following are the facts:

(1) p-Gauss $\Rightarrow$ p-Rademacher

(2) p-Rademacher $\Rightarrow$ $(p-\varepsilon)$-Gauss

(3) However, the type interval need not be open, i.e. the analogue for maps of Theorem 12.7 is false.

Example: Suppose $1 \leq p < 2$. Then L^p is *not* of type p-Gauss. For if it were, it would be of type $(p+\varepsilon)$-Rademacher, which is not the case. So the Gauss-type interval for L^p is $(0,p)$. (A direct, elementary proof of this is possible.)

Finally we mention an amazing theorem due to Maurey and Pisier.

Definition: E is of *infratype* p provided that, if $\left(\sum_n |x_n|^p\right)^{1/p} < +\infty$ then there is *some* choice of signs ε_n so that

$$\sum_n \varepsilon_n x_n \text{ converges} \quad .$$

Maurey/Pisier's result is that infratype p implies type $(p-\varepsilon)$ for $\varepsilon > 0$.

IV. Ultrapowers and Superproperties

Two Banach spaces E, F are said to be $(1+\varepsilon)$-*isometric* provided that there exists a linear bijection $u: E \to F$ such that $\|u\|\|u^{-1}\| \leq 1+\varepsilon$.

By scaling, we can arrange that

$$|x| \leq |u(x)| \leq (1+\varepsilon)|x| \quad \text{for all } x \in E.$$

We define a kind of distance between two spaces: $d(E,F) = \inf\{\|u\|\|u^{-1}\|: E \overset{u}{\to} F$ is an isomorphism$\}$. Of course $d(E,F) = +\infty$ if E and F are not isomorphic. If $d(E,F) = 1$ we say that E and F are *pseudo-isometric*.

The space F is *finitely representable in* E if, for every $\varepsilon > 0$, each finite dimensional subspace of F is $(1+\varepsilon)$-isometric to a subspace of E. Two spaces are called *pseudo-equivalent* if each is finitely representable in the other. This is obviously an equivalence relation.

Examples: (1) For fixed p, all infinite-dimensional L^p spaces are pseudo-equivalent. Indeed, any L^p space is finitely representable in ℓ^p, while ℓ^p is isometric to a subspace of any infinite dimensional L^p.

(2) Every Banach space X is finitely representable in ℓ^∞ (or c_0). For X is isometric to a subspace of $L^\infty(B^*)$, where B^* is the unit ball of X^*. When we are dealing with a finite dimensional subspace we can approximate B^* by finitely many points.

Consider P, a property of Banach spaces. We say that P is a *superproperty* provided that if E has P and F is finitely representable in E, then F has P.

Given any property P, there is a natural associated superproperty, "*super* P". We say that E has super P if and only if every F finitely representable in E has P. It is clear that super P is a superproperty, and super P is the "smallest" superproperty $\geq P$; that is, if Q is any sperproperty which implies P, then Q also implies super P.

Clearly if $p_1 \Rightarrow p_2$ then super $p_1 \Rightarrow$ super p_2.

Notation: It is convenient to write $F \partial\!\!\rightarrow G$ if F is finitely representable in G.

Examples of superproperties

(1) "dimension ≤ 3", though this is not a very interesting superproperty.

(2) "Type p" is a superproperty (whether Rademacher or Gauss type). The same is true of "cotype q". The reason is that what is involved is a set of finite dimensional inequalities

$$\left(E\left|\sum_n Z_n x_n\right|^q\right)^{1/q} \leq \tau_{qp}\left(\sum_n |x_n|^p\right)^{1/p} .$$

The "type constant" τ is itself a superproperty.

(3) Recall that E is "p-Pietsch" if every p-summing map from E is completely summing. We will see later that this is a superproperty.

(4) Let G_0 be a given Banach space. Then "$E \partial\!\!\rightarrow G_0$" is a superproperty of E. Likewise "$G_0 \partial\!\!\not\rightarrow E$" ("$G_0$ not finitely representable in E") is a superproperty of E.

(5) This leads to a class of very basic superproperties: "$L^p \not\!\!\partial\!\!\rightarrow E$". This means that E *fails* to contain $(1+\varepsilon)$-models of the finite dimensional ℓ^p_n space for large n.

In this connection, recall the fundamental

Dvoretzky Theorem: L^2 is finitely representable in every infinite dimensional Banach space. Thus, the property $L^2 \not\!\!\partial\!\!\rightarrow E$ is possessed by *no* infinite dimensional Banach space.

Given a Banach space E, it is of interest to determine the set of p for which L^p is finitely representable in E. We shall denote this set by S(E), the "*spectrum*" of E.

12.8. Proposition: S(E) is a closed subset of $[1,+\infty]$.

Proof: $p \in S(E)$ if and only if, for every n and every $\varepsilon > 0$, ℓ^p_n is $(1+\varepsilon)$-isometric to a subspace of E. It is obvious that this property extends to points p in the closure of S(E). ■

Additional facts about S(E)

(1) If E is infinite dimensional, $2 \in S(E)$ (Dvoretzky).

Note: There is a sharper formulation of the Dvoretzky Theorem: for every n and $\varepsilon > 0$ there exists an integer $N = N(n,\varepsilon)$ so that if E has dimension $\geq N$ then E has a subspace which is $(1+\varepsilon)$-isometric to ℓ_n^2. (The behavior of $N(n,\varepsilon)$ has been studied by Lindenstrauss and others.)

(2) $S(E) \cap [1,2]$ is a *closed interval*. For suppose $p < 2$ and $p \in S(E)$. If $p < q < 2$, one can embed L^q isometrically into L^p. (See the discussion using p-Gauss measure Γ_p in Lecture 10.) So $L^p \underset{\sigma}{\to} E$ implies that $L^q \underset{\sigma}{\to} E$.

(3) $S(E) \cap (2,\infty]$ seems to be rather "arbitrary".

The parts of $S(E)$ below and above 2 give rather different information about E.

A very important result is due to Maurey and Pisier; it is based on work of Pisier, Maurey, Krivine, and others.

12.9. Theorem (Maurey/Pisier): Suppose that $1 \leq p < 2$. Then E has p-Gauss type if and only if L^p is *not* finitely representable in E.

Remark: The "only if" part of this theorem is trivial. For "p-Gauss type" is a superproperty, and L^p does not have p-Gauss type.

Lecture 13. q-factorization, Maurey's Theorem Grothendieck Factorization Theorem

By virtue of Maurey/Pisier's Theorem (12.9) on Gauss types, we know that the type interval of E and the spectrum $S(E) \cap [1,2]$ are disjoint. We also know (by Theorem 12.7) that L^p is *not* of type p-Gauss for $1 \leq p < 2$. An important question: for a general Banach space E, is the type interval *exactly* the complement of $S(E) \cap [1,2]$? The answer is yes; this was first shown in 1977. More precisely:

13.1. Theorem: Let E be a Banach space. Suppose that $S(E) \cap [1,2] = [p_0,2]$ with $p_0 < 2$. Then the *type interval* of E is $[0,p_0)$. If $p_0 = 2$, the type interval may be either $[0,2)$ or $[0,2]$.

Remark: The type of a Banach space depends only on the topology, not the norm. Hence, by the theorem above, finite representability of L^p in E also is independent of the norm--a surprising conclusion.

We will eventually have something to say about the part of $S(E)$ above 2. But we turn now to another *factorization theorem.*

Suppose that $p < q$ and that $u: F \to L^p(T,dt)$ is a continuous linear map. We say that u is q-*factorizable* provided there exists a continuous linear map $v: F \to L^q(T,dt)$ and a function $h \in L^r(T,dt)$ (where $\frac{1}{r} = \frac{1}{p} - \frac{1}{q}$) so that the following diagram commutes:

$$\begin{array}{ccc} F & \xrightarrow{u} & L^p \\ & {\scriptstyle v}\searrow \quad \nearrow{\scriptstyle (h)} & \\ & L^q & \end{array}$$

If u is q-factorizable we define

$$\phi_{p,q}(u) = \inf\{\|h\|_{L^r}\|v\|\}$$

where the infimum is taken over all such factorizations of the map u.

13.2. Theorem (Maurey): The following condition is necessary and sufficient for u to be q-factorizable: there exists a constant ϕ_{pq} such that, for each finite sequence (f_n) in F,

$$\left(\int_T [\sum_i |u(f_i)(t)|^q]^{p/q} dt\right)^{1/p} \le \phi_{pq}\left(\sum_i |f_i|_F^q\right)^{1/q} .$$

The best constant ϕ_{pq} is $\phi_{pq}(u)$, as defined above.

Proof: The necessity of the condition follows by a straightforward application of Hölder's inequality. The proof of sufficiency is analogous to the proof of Pietsch's factorization theorem, but more elaborate.

13.3. Theorem: Suppose $u: F \to L^p$ and $p < q \leq 2$. Then u is q-factorizable if and only if u is of type q-Gauss.

Proof: Suppose that u is of type q-Gauss, and let (Z_i) be q-Gauss random variables. Then, by Theorem 12.4, there is a constant τ_{qp} such that

$$(*) \qquad \left(E_\omega\left|\sum_i u(f_i)Z_i\right|^p\right)^{1/p} \leq \tau_{qp}\left(\sum_i |f_i|^q\right)^{1/q} .$$

Here the left side involves a double integral in t and ω, namely

$$E_\omega\left(\int_T dt\left|\sum_i u(f_i)(t)Z_i(\omega)\right|^p\right) = \int_T dt\ E_\omega\left(\left|\sum_i u(f_i)(t)Z_i(\omega)\right|^p\right) .$$

But $\sum_i u(f_i)(t)Z_i$ is just a q-Gauss variable whose parameter is $\sigma = \left(\sum_i |u(f_i)(t)|^q\right)^{1/q}$. The expectation of the pth power $(p < q)$ is then $\sigma^p \|\gamma_q\|_p^p$. Accordingly inequality (*) becomes

$$\left(\int_T dt\left[\sum_i |u(f_i)(t)|^q\right]^{p/q}\right)^{1/p} \|\gamma_q\|_p \leq \tau_{qp}\left(\sum_i |f_i|^q\right)^{1/q} .$$

So, by Theorem 13.2, u is q-factorizable. The converse follows by reversing the argument. ■

Application: Grothendieck's Factorization Theorem

Suppose that $a \geq 2 \geq b \geq 0$ and that $u: L^a(X,\lambda) \to L^b(Y,\mu)$ is linear and continuous. Define r and s by the relations $\frac{1}{2} = \frac{1}{r} + \frac{1}{a}$, $\frac{1}{b} = \frac{1}{s} + \frac{1}{2}$. Then there exist functions $\alpha \in L^r(X,\lambda)$ and $\beta \in L^s(Y,\mu)$ and a continuous linear map $v: L^2(X,\lambda) \to L^2(Y,\mu)$ such that the following diagram commutes:

$$\begin{array}{ccc} L^a(X,\lambda) & \xrightarrow{\ u\ } & L^b(Y,\mu) \\ \downarrow{\scriptstyle(\alpha)} & & \uparrow{\scriptstyle(\beta)} \\ L^2(X,\lambda) & \xrightarrow{\ v\ } & L^2(Y,\mu) \end{array}$$

Proof (for $a < +\infty$; the case $a = +\infty$ will be discussed in Lecture 17): Since $\infty > a \geq 2$, L^a is of type 2. Hence the map $u: L^a \to L^b$ is of type 2. So by Theorem 13.3 u is 2-factorizable; there exists $w: L^a(X,\lambda) \to L^2(Y,\mu)$ and $\beta \in L^s(Y,\mu)$ so that $u = (\beta)\ w$:

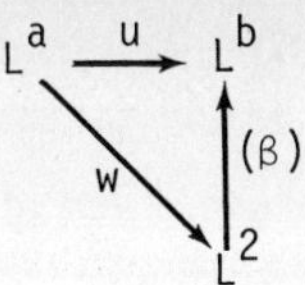

Now consider the transpose ${}^t w\colon L^2(Y,\mu) \to L^{a'}(X,\lambda)$. We can apply Theorem 13.3 again to deduce the existence of $z\colon L^2(Y,\mu) \to L^2(X,\lambda)$ and $\alpha \in L^r(X,\lambda)$ (note: $\frac{1}{2}+\frac{1}{r} = 1 - \frac{1}{a} = \frac{1}{a'}$) such that the following diagram commutes:

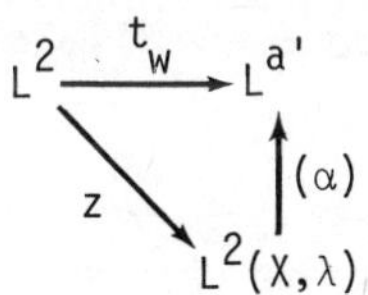

Now take $v = {}^t z\colon L^2(X,\lambda) \to L^2(Y,\mu)$. Then $w = {}^t z^t(\alpha) = v(\alpha)$ so that

$$u = (\beta)w = (\beta)\ v\ (\alpha). \quad ✱$$

Using the notion of q-factorizability we can deduce a number of properties equivalent to the q-Pietsch property. Recall that E is q-Pietsch provided that every q-summing map $u\colon E \to F$ is completely summing.

The Pietsch conjecture implies that every Banach space is $(1-\varepsilon)$-Pietsch. Another fact we shall eventually prove is that if E is q-Pietsch for $q > 2$ then E is finite dimensional. (Thus only the range $1 \leq q \leq 2$ is "interesting".)

Finally, we need to know that if p, q are fixed with $p < q$, and E is such that every q-summing map is p-summing, then E is q-Pietsch.

Lecture 14. Equivalent Properties, Summing vs. Factorization

14.1. Theorem: The following properties of a Banach space E are equivalent. Suppose $0 < p < q \leq 2$.

(1) Every continuous linear map $E' \to L^p$ factorizes as $E' \to L^q \xrightarrow{h} L^p$, where $h \in L^r$, $\frac{1}{r} = \frac{1}{p} - \frac{1}{q}$. (Equivalently, every map $E' \to L^p$ is of type Gauss q.)

(2) Every map $E' \to \ell^p$ factorizes as $E' \to \ell^q \xrightarrow{h} \ell^p$, $h \in \ell^r$ as in (1).

(3) Every scalarly ℓ^p sequence e in E can be written $e = hf$, i.e. $e_i = h_i f_i$, where $f \in S\ell^q(E)$ and $h \in \ell^r$, $\frac{1}{r} = \frac{1}{p} - \frac{1}{q}$.

(4) Every map from E into a Banach space which is q-summing must be p-summing.

Note: The implications (1) ⇒ (2) ⇒ (3) ⇒ (4) ⇒ (2) are valid for every Banach space E. For the implication (2) ⇒ (1) it is necessary to assume in addition either that E' has the metric approximation property or that $p \geq 1$, so that L^p has the metric approximation property.

Proof: (1) ⇒ (2) is trivial.

(2) ⇒ (3): For $e \in S\ell^p(E)$ define $u\colon E' \to \ell^p$, $u\colon \xi \mapsto \{\xi(e_n)\}_1^\infty$. The factorization of u in (2) gives the required factorization $e = hf$. (This also works for $S\ell^p(E'')$.)

(3) ⇒ (4): Suppose that $u\colon E \to F$ is q-summing. Let e be a scalarly ℓ^p sequence in E. We must show that $u(e)$ is an ℓ^p sequence in F. Now by (3), we write $e = hf$ with $h \in \ell^r$, $f \in S\ell^q(E)$; then $u(e) = hu(f)$. Since $u(f) \in \ell^q(F)$ it follows that $u(e) \in \ell^p(F)$, i.e. u is p-summing.

(4) ⇒ (2): (4) implies that there is a constant C such that, for every map $u\colon E \to$ Banach space,

$$\pi_p(u) \leq C\pi_q(u) \quad .$$

Turning to (2), it will be enough to get the stated factorization for maps defined by *finite* sequences, provided we establish a uniform bound. So let $(e''_k)_{k=1}^n$ be a finite sequence in E''. Define a map $e''\colon E' \to \ell_n^p \subset \ell^p$ by $(e''(\xi))_i = e''_i(\xi)$. Similarly, let $(\xi_i)_{i=1}^m$ be a sequence in E', and define the corresponding map $\Xi\colon E \to \ell_m^q$.

Now, by virtue of (4), there is a constant C such that $\pi_p(\Xi) \leq C\pi_q(\Xi)$. So for every sequence $e = (e_i)$ in E,

$$\begin{aligned} \|\Xi(e)\|_p &\leq \|e\|_p^* \pi_p(\Xi) \\ &\leq C\|e\|_p^* \pi_q(\Xi) \quad . \end{aligned}$$

Now in general, if a map v is r-summing, then v'' is r-summing with the same constant $\pi_r(v)$. Hence, for e'' a sequence in E'', we have

(*) $$\|\Xi(e'')\|_p \leq C\|e''\|_p^* \pi_q(\Xi) \quad .$$

At this point we need the formula

$$\|\Xi(e'')\|_p = \Big(\sum_k |\Xi(e_k'')|^p\Big)^{1/p}$$

$$= [\sum_k \Big(\sum_i | e_k''(\xi_i)|^q\Big)^{p/q}]^{1/p} \quad .$$

Next, we claim that $\pi_q(\Xi) \leq \Big(\sum_i |\xi_i|^q\Big)^{1/q}$. In other words, we assert that, for any sequence (e_j) in E,

$$\Big(\sum_j \|\Xi(e_j)\|^q_{\ell_n^q}\Big)^{1/q} \leq \Big(\sum_i |\xi_i|^q\Big)^{1/q} \|e\|_q^* \quad .$$

To see this, write out the left side as

$$\Big(\sum_{j,i} |\langle e_j, \xi_i \rangle|^q\Big)^{1/q} \quad .$$

Now the sum on j is

$$\sum_j |\langle e_j, \xi_i \rangle|^q \leq |\xi_i|^q (\|e\|_q^*)^q$$

so that

$$\sum_{ij} |\langle e_j, \xi_i \rangle|^q \leq \Big(\sum_i |\xi_i|^q\Big) \cdot (\|e\|_q^*)^q$$

which establishes our claim.

Thus inequality (*) becomes the inequality

(**) $$[\sum_k \Big(\sum_i \, e_k''(\xi_i)|^q\Big)^{p/q}]^{1/p} \leq C\|e''\|_p^* \Big(\sum_i |\xi_i|^q\Big)^{1/q} \quad .$$

Finally, what is the norm of e'' as a map from E' to ℓ_n^p? Clearly

$$\|e''\| = \sup_{|\xi| \leq 1} \Big(\sum_k |e_k''(\xi_k)|^p\Big)^{1/p}$$

$$= \|e''\|_p^* \quad .$$

Thus we may apply the Maurey factorization theorem (13.2) to get the desired conclusion (2), with the norm of the "factoring" function $h \in \ell^r$ bounded by a constant times $\|e''\|$.

(2) $\Rightarrow$ (1): We want to pass from ℓ^p to L^p. Here we assume that E' has the metric approximation property, or else that $p \geq 1$ (then L^p has the metric approximation property). Then we can deduce (1) by approximating with step functions. ¤

Remarks: 1. Property (4) says that E is q-Pietsch. Here we must assume the fact that (4) $\Rightarrow$ q-summing maps are completely summing, and also that q-Pietsch spaces cannot occur if $q > 2$ (unless the space is finite dimensional).

2. The following is another property equivalent to properties (1)-(4) above.

(5) (For $q > 1$) Every map $u: L^\infty \to E$ is q'-summing.

Here is a false (but "almost" correct) proof: Assume that $q > 1$ and E is q-Pietsch, i.e. every q-summing map on E is completely summing. Consider a map $u: L^\infty \to E$. Its transpose u' maps $E' \to L^1$ (not really). Now apply (1) above with $p = 1$. Thus u' factors as

$$E' \to L^q \xrightarrow{h} L^1 , \quad h \in L^{q'} .$$

Hence the original map u factors as

$$L \xrightarrow{h} L^{q'} \to E \quad .$$

Moreover we know that $L^\infty \xrightarrow{h} L^{q'}$ is q'-summing. So we have (almost) proved that u is q'-summing, i.e. that q-Pietsch $\Rightarrow$ (5). To fix this argument, note that it is enough to establish $\pi_{q'}(u) \leq C\|u\|$ on $\ell_n^\infty \subset L^\infty$, C independent of n. In other words, we reduce to finite dimensional approximations. Then the "transpose" argument above is valid, and the universal constant C comes from (4).

Now we turn to the converse. We show (5) $\Rightarrow$ (2) with $p = 1$. We begin with the special case $E = C(K)$ in the Pietsch factorization theorem--which perhaps we should have discussed at the time. $u: E \to F$ p-summing:

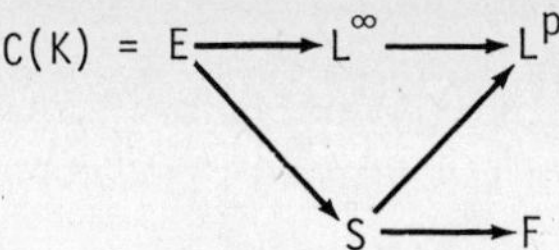

The unit ball of $C(K)^*$ is just the set of Radon measures μ on K with $\|\mu\| \le 1$. But in this special case we can arrange that the Pietsch measure is supported on K itself instead of this unit ball. In other words, there is a measure ν on K such that

$$|u(\phi)|_F \le \left(\int_k |\phi(k)|^p \nu(dk)\right)^{1/p} .$$

To see that this is so, we go back to the *proof* of the Pietsch Theorem. In the calculations we can take ξ to belong to the *extreme points* ($\cong K$) of the unit ball of $C(K)^*$.

Thus any p-summing map $u: C(K) \to F$ factors as

$$\begin{array}{ccc} C(K) & \longrightarrow & L^p(K,\nu) \\ & \searrow & \nearrow \\ & S & \longrightarrow F \end{array}$$

However, note that $C(K)$ is *dense* in $L^p(K,\nu)$. So we get a true factorization $C(K) \to L^p(K,\nu) \to F$.

Now to the proof that (5) ⇒ (2) with $p = 1$. Suppose that we have a map $u: E' \to \ell_n^1$. Its transpose maps $\ell_n^\infty \to E''$. Now we would rather have E here, not E''. To this end we invoke a theorem of Simmons (cf. Grothendieck's thesis): E'' is *always* finitely representable in E.

Thus, consider $C(K) = \ell_n^\infty \to E$ (K = finite set). By (5) this map is q'-summing, so it factors as

$$\ell_n^\infty \xrightarrow{h} \ell_n^{q'} \to E .$$

Here $h \in \ell_n^{q'}$ with $\|h\|_{q'} = 1$; $h^{q'}$ is the density of the probability measure ν. Taking the transpose again, we have the factorization of u as

$$E' \to \ell_n^q \xrightarrow{h} \ell_n^1 \ .$$

Thus (2) is satisfied with a bound independent of n. Accordingly we can pass to $n = \infty$: every map $E' \to \ell'$ factors through ℓ^q. ∎

3. Some further comments on the assertion that property (4) is the same as q-Pietsch. We claim that if E is such that every q-summing map is p-summing, with $p < q$ fixed, then q-summing maps are in fact (-1)-summing.

Why is this so? Consider the canonical map $E \to L^\infty(B',\lambda)$, where B' is the unit ball of E' with the weak* topology, and λ is a probability measure. Now the map

$$E \to L^\infty(B',\lambda) \to L^q(B',\lambda)$$

is certainly q-summing. Hence by (4) it is p-summing. So it has a Pietsch factorization; there is a Pietsch measure μ, also a probability measure on B', such that we have the factorization

$$\begin{array}{ccccc} E & \longrightarrow & L^\infty(B',\lambda) & \longrightarrow & L^q(B',\lambda) \\ & \searrow & & \nearrow & \\ & & L^p(B',\mu) & & \end{array}$$

That is, for every probability measure λ on B', there is a probability measure μ on B' such that, for all $x \in E$,

$$\left[\int |\langle x,\xi\rangle|^q d\lambda(\xi)\right]^{1/q} \le C\left[\int |\langle x,\xi\rangle|^p d\mu(\xi)\right]^{1/p}$$

where C is some universal constant.

Note that we can assume that $\mu \ge \frac{1}{2}\lambda_0$ simply by replacing μ by $\frac{1}{2}(\mu+\lambda_0)$ (and modifying C), where λ_0 is a fixed probability measure.

Then we may apply the Kakutani fixed-point theorem. Namely, given $\lambda \ge \frac{1}{2}\lambda_0$, let $F(\lambda)$ be $\{\mu \ge \frac{1}{2}\lambda_0$: μ is a probability measure such that the above inequality holds with the Pietsch constant $C\}$. Then $F(\lambda)$ is a compact, convex set. Moreover the multiple-valued mapping F has a closed graph. So, according to Kakutani, F

has a fixed point, i.e. there exists a probability measure $\lambda \geq \frac{1}{2}\lambda_0$ such that $\lambda \in F(\lambda)$.

This means that the $L^q(\lambda)$ and $L^p(\lambda)$ norms are equivalent on the subspace of functions $\xi \mapsto \langle x,\xi\rangle$, $x \in E$. We want to conclude that on this subspace *all* the L^r topologies are equivalent, and in fact they are equivalent to the "L^0 topology" --the topology of convergence in measure. (Thus, in our factorization, we will be able to replace p by 0, and finally by -1.)

Lecture 15. Non-existence of $(2+\varepsilon)$-Pietsch Spaces, Ultrapowers

15.1. Lemma: Let $E \subseteq L^0$ be a linear subspace on which the L^p and L^q topologies are equivalent, with $p < q$. Then the L^q topology is equivalent to the L^r topology for *all* $r < q$, including $r = 0$.

Proof: Clearly L^q convergence implies L^r convergence. Conversely, arguing by contradiction, suppose that L^0 convergence does *not* imply L^q convergence-- i.e. suppose there is a sequence (f_n) in E with $f_n \to 0$ in L^0 but not in L^q. If the norms $\|f_n\|_q$ are bounded by a constant C this is impossible, for on the *unit ball* of L^q it is easy to see that the L^0 and L^p topologies coincide; this is a trivial consequence of Holder's inequality. Hence $\|f_n\|_q \to \infty$. So let $g_n = f_n/\|f_n\|_q$. Then $g_n \in E$ and *a fortiori* $g_n \to 0$ in L^0. Moreover $\|g_n\|_q = 1$. Hence $\|g_n\|_p \to 0$. But since $g_n \in E$ it follows that $\|g_n\|_q \to 0$, a contradiction.∎

Now, returning to the set-up at the end of Lecture 14, we see that for every probability measure λ on B' there is a probability measure $\mu \geq \frac{1}{2}\lambda$ such that the $L^q(\mu)$ and $L^p(\mu)$ topologies are equivalent on $E \subset L^0(B',\mu)$. Hence μ is a Pietsch measure for *all* L^r, $0 < r < q$. And therefore the map $E \to L^q(B',\lambda)$ is r-summing for all $0 < r < q$.

Accordingly, every q-summing map $u: E \to F$ is actually r-summing, $0 < r < q$. E.g. take $r = \frac{1}{2}$. Then u is *completely* summing, by the Pietsch conjecture (8.3).

Our arguments have established the following (cf. 14.1).

15.2. Theorem: The following are equivalent:

(1) E is q-Pietsch, $0 < q \leq 2$.

(2) For some $p < q$ (or *all* such p), every map $E' \to L^p$ is q-factorizable or of type q-Gauss. N.B. Here we need to assume either that $p > 1$ or that E' has the metric approximation property.

(2') For some (all) $p < q$, with no additional assumptions, every map $E' \to \ell^q$ is q-factorizable or of type q-Gauss. Equivalently, every $S\ell^p$ sequence in E is the product of a $S\ell^q$ sequence and a real ℓ^r sequence, where $\frac{1}{p} = \frac{1}{q} + \frac{1}{r}$.

(3) Every map $L^\infty \to E$ (or $C(K) \to E$) is q'-summing, where $\frac{1}{q} + \frac{1}{q'} = 1$.

Note: The proof of (3) uses the finite representability of E'' in E.

This result has some interesting consequences. First of all, we can show that if $q > 2$ there are no non-trivial q-Pietsch spaces.

15.3. Theorem: If E is infinite dimensional, then E cannot be $(2+\varepsilon)$-Pietsch, $\varepsilon > 0$.

Proof: First we show that the q-Pietsch property is a *superproperty*. The map $E \to L^\infty \to L^q$ is supposed to be completely summing. So $p < q$ implies $\pi_p(u) \leq C\pi_q(u)$, where C is some universal constant. Now suppose that F is finitely representable in E. Consider a map $F \to L^\infty \to L^q$. We may assume that F is finite dimensional, so that $F \hookrightarrow E$ almost isometrically. Hence by Nachbin's theorem the map $F \to L^\infty$ extends to $E \to L^\infty$ with almost the same norm:

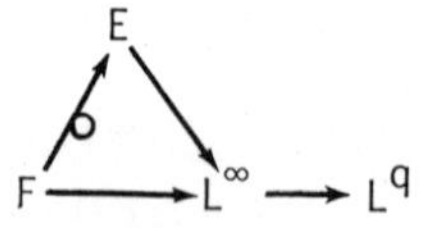

It follows that the map $F \to L^\infty \to L^q$ is completely summing. Hence we get a π_p, π_q estimate for F with constant $\leq C(1+\varepsilon)$. This uniform estimate extends to the general, infinite-dimensional, case.

Now Hilbert space ℓ^2 is finitely representable in every infinite-dimensional Banach space. So if there were an infinite-dimensional q-Pietsch space, $q > 2$, then ℓ^2 would be q-Pietsch. We will show this is not the case. Using property (3) of Theorem 15.2 we will exhibit a map $\ell^\infty \xrightarrow{\alpha} \ell^2$ which is *not* q'-summing, $q' < 2$. Here $\alpha \in \ell^2$ is a multiplication operator.

Consider the basis vectors $\{e_n\}_1^\infty$ in ℓ^∞ (or, better, c_0). This sequence is scalarly ℓ^1 since $c_0^* = \ell^1$. Hence it is also scalarly $\ell^{q'}$. Now if (α) is q'-summing, the image $\{\alpha_n e_n\}_1^\infty$ should be an $\ell^{q'}$-sequence. That is, $\sum_n |\alpha_n|^{q'}$ should converge. But this *need not* be the case, since $q' < 2$. Thus ℓ^2 is *not* q-Pietsch. ⌘

The proof of Theorem 15.3 exemplifies a useful general strategy. To show that a superproperty is *never* verified, it suffices to show that Hilbert space doesn't have it.

It is also worth noting that Theorem 15.2 yields a *new proof of the Pietsch conjecture*. For E' type q-Gauss implies that E is q-Pietsch. But every Banach space E' is of type 1-Rademacher, hence type $(1-\varepsilon)$-Gauss. And therefore E is type $(1-\varepsilon)$-Pietsch. But this is just a restatement of the Pietsch conjecture.

A few more remarks: That E' is of Gauss type p implies that E is p-Pietsch, but the converse is false. (Example: L^1 is 2-Pietsch while L^∞ is of type only 1.) Note, however, that if E' is of Gauss type p then so is every *subspace* of E'; therefore every *factor space* of E is p-Pietsch. (The p-Pietsch property does not automatically pass to factor spaces in general; remember that L^1 is 2-Pietsch, but not all its factor spaces, since every Banach space is a factor space of L^1.) Maurey/Pisier proved the converse: for $p < 2$, if *all the factor spaces* of E are p-Pietsch, then E' is of Gauss-type p.

Here is a further result of Pisier.

15.4. <u>Theorem</u>: E is type 1-Gauss ⇔ E' is also.

<u>Proof</u>: It is enough to prove the implication ⇒; for then if E' is type 1 so is E", and $E \subseteq E"$. Now we know (from Theorems 12.9 and 13.1) that a space is of

type 1-Gauss if and only if L^1 is *not* finitely representable in it. So assume that E' is not type 1. Then $L^1 \underset{\sigma}{\to} E'$. This is "practically" the same as saying that L^1 is embedded in E'. Let's pretend that in fact $L^1 \subseteq E'$. Then L^∞ is a quotient of E". We have

$$L^\infty \cong E''/N$$

where N is a weak* closed subspace of E".

Now recall Nachbin's theorem: if $F \subseteq G$ and $u: G \to L^\infty$, we can extend u to G. Dualizing this, and using the fact that N is weak* closed, we can lift the map $L^1 \hookrightarrow L^\infty = E''/N$ to an embedding $L^1 \hookrightarrow E''$. (L^1 embeds in L^∞ since every Banach space embeds in L^∞.)

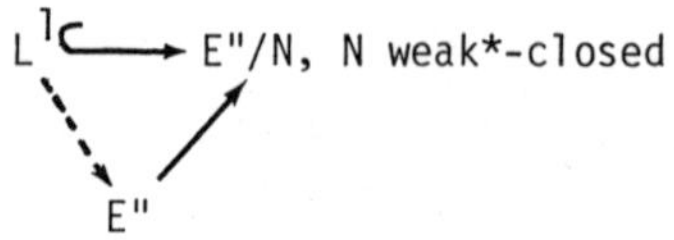

$$\begin{array}{ccc} L^1 & \hookrightarrow & E''/N, \text{ N weak*-closed} \\ & \searrow \quad \nearrow & \\ & E'' & \end{array}$$

More careful argument along these lines shows that if L^1 is finitely represented in E' then it is also finitely represented in E". Since $E'' \underset{\sigma}{\to} E$ we end up with $L^1 \underset{\sigma}{\to} E$, so that E is not of type 1 if E' is not of type 1. ∎

Remark: The result above does *not* extend to exponents q not equal to 1.

Ultrapowers

Let $(E_i)_{i \in I}$ be an indexed family of Banach spaces and let $\mathcal{U}$ be a non-trivial ultrafilter on the index set I. We will define the *ultraproduct* $\Pi_{i \in I} E_i/\mathcal{U}$ to be the following quotient space.

Let $\ell^\infty(E)$ be the space of bounded sequences in E. Then, more generally, by $\ell^\infty((E_i)_{i \in I})$ we mean the families $(e_i)_{i \in I}$ with $e_i \in E_i$ and $\sup_{i \in I} \|e_i\| < \infty$. Now let N be the null sequences: $\|e_i\| \underset{\mathcal{U}}{\to} 0$ along the ultrafiler $\mathcal{U}$. The quotient

$$\ell^\infty((E_i)_{i \in I})/N$$

is by definition the *ultraproduct* of the E_i relative to $\mathcal{U}$. Since N is obviously closed, this is a Banach space.

If all the E_i coincide with E, the ultraproduct is an *ultrapower* of E. Some facts about ultrapowers:

15.5. Theorem: An ultrapower of an ultrapower of E is an ultrapower of E. (Perhaps surprisingly, this is *not* trivial to prove.)

15.6. Theorem: E is embedded in any ultrapower of E. Also, E is embedded in the ultraproduct of all finite dimensional subspaces of E, indexed by themselves.

Proof: To $x \in E$ assign $\{x_F: F$ a finite-dimensional subspace of $E\}$, where $x_F = x$ as soon as $x \in F$ (and x_F is arbitrary otherwise). This induces an isometric linear embedding of E in the ultraproduct. ■

The next result gives the real reason for our interest in ultrapowers.

15.7. Theorem: F is finitely representable in $E \Leftrightarrow F$ is embedded in some ultrapower $E^I/\mathcal{U}$.

Proof: Suppose $F \xrightarrow[\sigma]{} E$. Let $(F_i)_{i\in I}$ be the self-indexed family of finite dimensional subspaces of F. Then for each i and each $\varepsilon > 0$ there is a $(1+\varepsilon)$-isometry $u_{i,\varepsilon}: F_i \to E$. So take as our final index set J the pairs (i,ε). We must construct a suitable ultraproduct over this index set. Let $\mathcal{U}$ be an ultrafilter on J finer than the filter $F_i \uparrow F$, $\varepsilon \downarrow 0$.

If $x \in F$, then eventually $u_{i,c}(x)$ makes sense, since eventually $x \in F_i$. So define $u(x)$ to be the image of $(u_{i,\varepsilon}(x))_{i,\varepsilon}$ in $E^J/\mathcal{U}$. Clearly u is linear, and u is an isometry because $\|u_{i,\varepsilon}(x)\| \to \|x\|$ as $(i,\varepsilon) \to \infty$ along $\mathcal{U}$. Thus F is embedded in the ultrapower $E^J/\mathcal{U}$.

Conversely, suppose F is embedded in an ultrapower $E^J/\mathcal{U}$. To prove that F is finitely representable in E it suffices to show that $E^J/\mathcal{U}$ is finitely representable in E. So consider a finite dimensional subspace $F \subseteq E^J/\mathcal{U}$ with a basis $f_1, f_2, \ldots, f_n$. Let $(f_{1,i}), \ldots, (f_{n,i})$ be families in E^J representing $f_1, \ldots, f_n$ respectively. For each i define $u_i: F \to E$ to be the map sending $\lambda_1 f_1 + \cdots + \lambda_n f_n$ to $\lambda_1 f_{1,i} + \cdots + \lambda_n f_{n,i}$. We can show that this approaches an isometry as $i \to \infty$. Indeed, compare

$$\|\lambda_1 f_1 + \cdots + \lambda_n f_n\| \text{ with } \|\lambda_1 f_{1,i} + \cdots + \lambda_n f_{n,i}\| .$$

Clearly the latter converges to the former as $i \to \infty$. Thus we have a family of norms $\|\cdot\|_i$ on F such that, for $x \in F$, $\|x\|_i \to \|x\|$ as $i \to \infty$. This implies that, for i sufficiently large, $\|\cdot\|_i \leq C\|\cdot\|$, C some constant. So the family of norms is eventually equicontinuous.

Then Ascoli's theorem says that $\|x\|_i \to \|x\|$ uniformly for x in the unit ball of F, a compact set. Hence, given $\varepsilon > 0$, if i is sufficiently large the map u_i is a $(1+\varepsilon)$-isometry of F into E. ∎

As an application, suppose that P is a *topological* vector space property of E. Then "super-P" is also a topological property. (Example: super-reflexivity, a very interesting property.)

By a topological property P of Banach spaces, we mean one which is preserved under linear topological isomorphism.

Suppose now that E_1 and E_2 are isomorphic, and that E_1 has super-P, i.e. all the ultrapowers of E_1 have P. The isomorphism $E_1 \simeq E_2$ induces an isomorphism of corresponding ultrapowers. Therefore E_2 also has super-P.

Examples of topological super-properties: "type p", "q-Pietsch".

We also mention the deep theorem that every ultraproduct of L^p spaces is an L^p space (over some huge measure space). Hence F is finitely representable in L^p $\Leftrightarrow$ F is embedded in some huge L^p. Thus "F embedded in some L^p space" is a super-property.

Lecture 16. The Pietsch Interval. The Weakest Non-trivial Superproperty. Cotypes, Rademacher vs. Gauss

Let E be a Banach space. We define the *Pietsch interval* of E to be the set $(-1,q_0|$ of q such that E is q-Pietsch. By Theorem 15.3, this interval must be contained in $(-1,2]$.

The following result of H. Rosenthal (1971) characterizes the possibilities.

16.1. Theorem: The Pietsch interval of E must be either $(-1,2]$ or $(-1,2)$ or $(-1,q_0)$ for some $q_0 < 2$. In other words, if $q < 2$ and E is q-Pietsch, then E is $(q+\varepsilon)$-Pietsch for some $\varepsilon > 0$.

This theorem is hard, though not so difficult as the corresponding result for the *type interval*, Theorem 12.7. Of special interest is the observation that every Banach space E is $(1-\varepsilon)$-Pietsch; but if E is 1-Pietsch, then it is also $(1+\varepsilon)$-Pietsch.

Recall that if $q > 1$ then E is q-Pietsch if and only if every map $L^\infty \to E$ (or $C \to E$) is q'-summing. Now we can also give a criterion for E to be 1-Pietsch--namely, that E be q-Pietsch for some $q > 1$, or that there exist $q' < \infty$ so that every map $L^\infty \to E$ is q'-summing.

We also mention a theorem, proved in 1977 by Maurey, Pisier and Krivine, concerning $S(E) = \{p: L^p$ is finitely representable in E$\}$. We know that $S(E) \cap [1,2]$ is a closed interval, the complement of the interval of types. Not much is known about $S(E) \cap [2,\infty)$ except that it is, of course, a closed set. Perhaps an arbitrary closed set can occur. There are examples for which it is finite; thus $S(L^r) \cap [2,\infty) = \{2,r\}$. By taking direct sums one can get arbitrary finite sets.

16.2. Theorem: Let q' (possibly $+\infty$) be the largest element of S(E). Then $q' = (q_0)'$, the index conjugate to the supremum q_0 of the Pietsch interval.

It follows that E is 1-Pietsch if and only if L^∞ is not finitely representable in E. For E is 1-Pietsch $\Leftrightarrow q_0 > 1 \Leftrightarrow q_0' < \infty \Leftrightarrow L^\infty$ is not finitely representable in E.

These results lead to a classification of the *hierarchy of superproperties* for infinite dimensional Banach spaces. There is, to begin with, a *strongest* superproperty, namely H(E): "E is a Hilbert space". For, as we have noted, if P is any superproperty which has an infinite dimensional example, then $H(E) \Rightarrow P(E)$, i.e. P is true for Hilbert spaces. Also, there is a *weakest* superproperty W(E): "E is an infinite-dimensional Banach space".

The interesting thing is that after W there is a *first* non-trivial super-property F(E): "L^∞ is not finitely representable in E". In other words, if P is a superproperty which is not universally true, then P fails for L^∞. For every Banach space is finitely representable in L^∞. Now the property F is equivalent to "1-Pietsch". Thus "1-Pietsch" is the weakest non-trivial superproperty. So if E has *any* superproperty (e.g. super-reflexivity, type p) not common to all Banach spaces, then E is 1-Pietsch, hence $(1+\varepsilon)$-Pietsch.

Suppose that E is q-Pietsch. Take any $p < q$. Then we know that every sequence e in $S\ell^p(E)$ is of the form αf where $\alpha \in \ell^r$ and $f \in S\ell^q(E)$, $\frac{1}{r} = \frac{1}{p} - \frac{1}{q}$. If $q > 1$ we may take $p = 1$. So $e \in S\ell^1(E)$ implies that $e = \alpha f$, $f \in S\ell^q(E)$, $\alpha \in \ell^{q'}$. In particular, it follows that if $e \in S\ell^1(E)$ then $e \in \ell^{q'}(E)$.

All this is an old problem. In particular, Dvoretzky and Rogers observed that in general $S\ell^1$ sequences are not ℓ^1. But for the case of Hilbert space, $S\ell^1 \Rightarrow \ell^2$. From our point of view this follows because a Hilbert space is 2-Pietsch. We now have achieved a generalization of this fact.

In the converse direction we have the following.

16.3. <u>Proposition</u>: Suppose that $S\ell^1(E) \subseteq \ell^{q'-\varepsilon}(E)$ for some $\varepsilon > 0$. Then there exists $\delta > 0$ such that E is $(q+\delta)$-Pietsch.

In particular, E is 1-Pietsch $\Leftrightarrow$ $S\ell^1(E) \subseteq \ell^{q'}(E)$ for some $q' < \infty$. (So this is another form of the weakest nontrivial superproperty.)

<u>Proof</u>: Consequence of Theorem 15.2. ∎

Now we will give yet another version of the first nontrivial superproperty. Let (ε_n) be Rademacher random variables and (Z_n) 2-Gauss R.V.'s. Consider the series $\sum_n \varepsilon_n x_n$, $\sum_n Z_n x_n$. From the results in Lecture 12, one sees that if the second series converges a.s., so does the first. But the converse is false.

<u>Example</u>: Take $E = c_0$ (since we're dealing with a superproperty, this is the place to look for a counterexample). Let (η_n) be the standard basis of c_0. Then

the series $\sum_n \varepsilon_n\alpha_n\eta_n$ converges in c_0 $\Leftrightarrow$ $\alpha_n\varepsilon_n \to 0$ $\Leftrightarrow$ $\alpha_n \to 0$. On the other hand, consider the series $\sum Z_n\alpha_n\eta_n$. This converges $\Leftrightarrow$ $Z_n\alpha_n \to 0$. Now it is easy to find sequences $\alpha_n \to 0$ such that $Z_n\alpha_n$ does *not* converge a.s. to 0--simply because the R.V.'s Z_n are unbounded.

In 1973 Maurey/Pisier proved that a Banach space E has the property that $\sum\varepsilon_n x_n \to 0$ a.s. $\Rightarrow$ $\sum Z_n x_n \to 0$ a.s. if and only if E has the first nontrivial superproperty (ℓ^∞ not finitely representable in E).

Cotypes

Definition: E is *cotype* q-*Rademacher* provided that if (x_n) is a sequence in E then

$$\sum_n \varepsilon_n x_n \text{ converges a.s.} \Rightarrow \left(\sum_n |x_n|^q\right)^{1/q} < +\infty \quad .$$

Note that this is only interesting for $2 \leq q \leq +\infty$, since the case $E = \mathbb{R}$ shows that one cannot expect that $q < 2$.

There is also a notion of *Gauss cotype*. Here we need $q \leq 2$, and in fact only $q = 2$ is interesting.

16.4. Proposition: E is cotype 2-Rademacher $\Leftrightarrow$ cotype 2-Gauss.

Proof: Since $\|Z_n\|_1 \geq a > 0$, if $\sum_n Z_n x_n$ converges a.s. so does $\sum_n \varepsilon_n x_n$. Hence if E is of cotype 2-Rademacher, it is of cotype 2-Gauss.

Conversely, assume that E is of cotype 2-Gauss. Now "cotype 2-Gauss" is a superproperty. (For, via results of Lecture 12, there are universal "geometric" constants associated with this property, as with "type p".) Moreover the space c_0 does *not* have this property. Hence "cotype 2-Gauss" $\Rightarrow$ "weakest nontrivial superproperty", which, by Maurey and Pisier's 1973 result quoted above, implies that the series $\sum\varepsilon_n x_n$ and $\sum Z_n x_n$ are simultaneously a.s. convergent. Hence E has cotype 2-Rademacher. ∎

The above is a "fantastic" argument, since it turns on the structure of the superproperty hierarchy rather than on the detailed property in question.

Every Banach space has cotype $+\infty$. The interval of cotypes $(q_0',+\infty]$ or $[q_0',+\infty]$ may be open or closed (like the Rademacher type interval).

16.5. <u>Theorem</u>: The infimum q_0' of the cotype interval is the conjugate of q_0, the supremum of the Pietsch interval.

If the Pietsch interval is $(-1,2|$ then E has cotype 2; or else cotype $2+\varepsilon$ for every $\varepsilon > 0$. If the Pietsch interval is $(-1,1)$ then E has cotype $+\infty$ only. Thus, having a finite cotype is equivalent to the weakest nontrivial super-property.

<u>Relations between Type and Cotype of a Banach Space</u>

16.6. <u>Proposition</u>: If E has type Rademacher-p then E' has cotype Rademacher-p'.

<u>Proof</u>: Note that the dual of $\ell^p(E)$ is $\ell^{p'}(E')$. Now take a sequence (ξ_i) in $\ell^{p'}(E')$. Then, given $\varepsilon > 0$, there is a sequence (x_i) in $\ell^p(E)$ such that

$$(1-\varepsilon)\left(\sum_i |x_i|^p\right)^{1/p}\left(\sum_i |\xi_i|^{p'}\right)^{1/p'} \le |\sum_i \langle x_i,\xi_i\rangle| \quad .$$

(We will ignore ε in the rest of the argument.)

Now

$$|\sum_i \langle x_i,\xi_i\rangle| = |E_\varepsilon\{\sum_{i,j} \langle \varepsilon_i x_i,\varepsilon_j\xi_j\rangle\}|$$

$$\le [E_\varepsilon|\sum_i \varepsilon_i x_i|^p]^{1/p}[E_\varepsilon|\sum_j \varepsilon_j\varepsilon_j|^{p'}]^{1/p'} \quad .$$

Since E has type p, the latter is

$$\le C[\sum_i |x_i|^p]^{1/p}[E_\varepsilon|\sum_j \varepsilon_j\xi_j|^{p'}]^{1/p'} \quad .$$

Hence

$$[\sum|\xi_i|^{p'}]^{1/p'} \le C[E_\varepsilon|\sum_j \varepsilon_j\xi_j|^{p'}]^{1/p'} \quad .$$

But the existence of an estimate of this form is equivalent to cotype p'. (Here, as elsewhere, we are using Kahane's inequality 11.1.) ∎

Examples: (a) For $2 \leq r \leq +\infty$, L^r is the dual of $L^{r'}$, with $1 \leq r' \leq 2$. Thus L^r has cotype r. Moreover it is not hard to see that L^r is not of any smaller cotype, so the cotype interval of L^r is $[r,+\infty]$.

(b) For $1 < r \leq 2$, L^r has cotype 2. Indeed, in this case $(L^r)' = L^{r'}$ has type 2.

(c) Perhaps surprisingly, the *converse* of Proposition 16.6 is *false*. In particular, L^1 has cotype 2, although L^∞ is very far indeed from having type 2 (L^∞ has only the trivial superproperty.)

Cotype, unlike type, is inherited by subspaces but *not* by quotients. It is easy to see what the difficulty is. Suppose E has type p and (x_n) is a sequence in E/N with $\sum|x_n|^p < +\infty$. The sequence (x_n) can be lifted to a sequence in E with the same property. The latter sequence must then converge a.s., so $\sum \varepsilon_n x_n$ converges a.s. in E/N. On the other hand, when we are dealing with cotypes, we run into difficulty. One cannot lift a.s. convergence from E/N to E.

Now every Banach space is a quotient of L^1, and L^1 has cotype 2. This shows concretely the failure of quotients to inherit cotype.

Summary

Let E be an infinite dimensional Banach space. Then $1 \leq p_0 \leq 2$, $2 \leq q_0 \leq \infty$, where p_0 and q_0 satisfy:

(1) $p_0 = \min\{p: L^p$ is finitely represented in E$\}$ (closed set)

$= \sup\{p:$ E is of type p$\}$

(2) $q_0 = \max\{q: L^q$ is finitely represented in E$\}$

$= \inf\{q:$ E is of cotype q$\}$

$= \inf\{q:$ every $S\ell^1$ sequence in E is $\ell^q\}$

(3) $q_0' = \sup\{p:$ E is p-Pietsch$\}$.

Further Remarks

1. It is easy to see from the facts above and the definitions that $q_0(E') \leq p_0(E)'$. One can have strict inequality here. Thus if $E = c_0$, $p_0(E) = 1$ so $p_0(E)' = +\infty$; but $E' = \ell^1$ has cotype 2, and $2 < \infty$.

2. If E is of type p-Rademacher, then E' is of cotype p'-Rademacher. Hence, if $1 < r \leq 2$, L^r is of cotype 2; if $2 \leq r \leq +\infty$, L^r is of cotype r.

3. If $p \leq 1$ then the space L^p is of cotype 2. For the very important case of L^1 there is an elementary proof, as follows. (Cf. the proof of Theorem 11.5.)

Suppose L^1 is $L^1(T,dt)$. Then

$$E_\omega \|\sum_n \varepsilon_n(\omega)x_n\|_{L^1(T,dt)} = E_\omega E_t\left\{|\sum_n \varepsilon_n(\omega)x_n(t)|\right\}$$

$$= E_t E_\omega\left\{|\sum_n \varepsilon_n(\omega)x_n(t)|\right\} .$$

Now, by the fact that $\mathbb{R}$ is of cotype 2, or by Khintchine's inequality, this is

$$\geq C\int_T \left\{\sum_n |x_n(t)|^2\right\}^{1/2} dt$$

$$= C\|x\|_{L^1(\ell^2)}$$

$$\geq C\|x\|_{\ell^2(L^1)} = C\left(\sum_n \|x_n\|^2_{L^1}\right)^{1/2} .$$

Thus L^1 has cotype 2. ∎

Lecture 17. Gauss-summing Maps. Completion of Grothendieck Factorization Theorem. TLC and ILL

Consequences of Cotype 2

We first discuss *Gauss-summing maps*.

Suppose that μ is a cylindrical probability on E, and $u: E \to F$ is a linear map. Recall that u is p-*Radonifying* provided $\|u(\mu)\|_p \leq \pi_p(u)\|\mu\|_p^*$ for all such μ.

Suppose now that we restrict our attention to *Gauss cylindrical probabilities* μ. Here the best definition is that μ is induced from a Gaussian probability on Hilbert space H via a continuous linear map $H \to E''$. In particular this means that the finite-dimensional sections of μ are Gaussian.

Definition: $u\colon E \to F$ is *Gauss-summing* provided that for all Gauss cylindrical probabilities μ on E, $u(\mu)$ is a *Radon* probability on $\sigma(F'',F')$.

Then there must exist a constant $\gamma_p(u)$ such that

$$\|u(\mu)\|_p \leq \gamma_p(u)\|\mu\|_p^* \quad .$$

(This relies on the Shepp-Landau-Fernique theorem, Lecture 7.) Of course, $\gamma_p(u) \leq \pi_p(u)$; the latter might be $+\infty$.

The Gauss-summing property can also be expressed using only *Radon* Gauss measures μ on finite dimensional subspaces. The condition is that, for some p, there is a constant $\gamma_p(u)$ such that the inequality holds for all such μ. One is usually interested in the case: $p = 2$.

Now a finite dimensional subspace is the image of $\mathbb{R}^n$, so if (e_n) is a suitable basis, and we write a random vector as $\sum e_n Z_n$, then the Z_n are independent Gaussian random variables. So the Gauss summing property for u tells us that

$$\left(E[\|\sum u(e_n)Z_n\|_F^2]\right)^{1/2} \leq \gamma_2(u)\|(e_n)\|_2^* \quad .$$

Indeed, the scalar order of $\sum e_n Z_n$ is given by

$$\sup_{|\xi|\leq 1} \left(E|\sum_n \langle \xi, e_n\rangle Z_n|^2\right)^{1/2} = \sup_{|\xi|\leq 1} \left(\sum_n |\langle \xi, e_n\rangle|^2\right)^{1/2} = \|(e_n)\|_2^* \quad .$$

Since $\|(e_n)\|_2^* \leq \|(e_n)\|_2$, the inequality above (with the universal constant $\gamma_2(u)$) implies that u is of type 2.

Thus we have the following *scale* of types of maps: completely summing $\Rightarrow$ p-summing for a finite $p \Rightarrow$ Gauss-summing $\Rightarrow$ type $2 \Rightarrow$ type k, $1 < k < 2 \Rightarrow$ type Gauss-1.

Remark: A Banach space E has type p provided the identity I has type 1. But there is no analogue for p-summing: I is never p-summing (unless E is finite dimensional). Similarly, I is never Gauss-summing unless E is finite dimensional. The proof is as follows:

The Gauss-summing property can be expressed by an inequality of the type

$$\{E|\sum_n e_n Z_n|^2\}^{1/2} \leq C\|e\|_2^* \ .$$

Accordingly, it is a superproperty. So if any infinite dimensional Banach space E has this property, Hilbert space has it. However that is not the case.

Thus if E is an infinite dimensional Banach space it must carry a cylindrical Gauss probability which is not Radon. (It might be interesting to try to prove this *without* the use of the Dvoretzky-Rogers theorem.)

Now we turn to two important consequences of the cotype 2 property.

17.1. Theorem: Suppose $u: E \to F$ is a linear map, u is Gauss-summing, and F has cotype 2. Then u is 2-summing.

Proof: Suppose e is a sequence with $\|e\|_2^* < \infty$. Now since F is of cotype 2,

$$\begin{aligned}\|u(e)\|_2 &\leq C\{E|\sum u(e_n)Z_n|^2\}^{1/2}\\ &\leq C'\|e\|_2^*\end{aligned}$$

Thus u is 2-summing. ∎

17.2. Theorem: Let $u: E \to F$ be a linear map, with E of cotype 2. Then if u is 2-summing, it is completely summing. That is, cotype 2 ⇒ 2-Pietsch.

Proof (sketch): To show that E is 2-Pietsch, we use the criterion of Theorem 15.2. We must show that any map $L^\infty \to E$ is 2-summing, i.e. that it factors through a suitable L^2 space.

Here we need an analogue of Theorem 13.3. Maurey has proved the following:

If $a < \infty$ and $L^a(x,\lambda) \overset{u}{\to} E$ is a map into a Banach space of cotype 2, then there exists a factorization of u as:

$$L^a \overset{(\alpha)}{\to} L^2(x,\lambda) \overset{v}{\to} E \ ,$$

where (α) is multiplication by $\alpha \in L^r(x,\lambda)$, $1/2 = 1/a + 1/r$.

This is a false for $a = +\infty$. Counter-example: take $E = \mathbb{R}$; u is just a continuous linear functional on L^∞. If it factored as above, then, since v is given by integration against an L^2 function ϕ, u would be given by $u(f) = \int \phi\alpha f d\lambda = \int g f d\lambda$ where $g = \phi\alpha \in L^1$. But the dual of L^∞ is not L^1!

What remains true, but has to be proved by other means (cf. pages 64-65) is the following: Any map $C(K) \overset{u}{\to} E$, E of cotype 2, factors as

$$C(K) \overset{(\alpha)}{\to} L^2(K,\lambda) \to E$$

where $\alpha \in L^2(K,\lambda)$ and λ is a suitable probability measure on K. (This measure depends on u and is not arbitrary.) So any map from $C(K)$ to E is 2-summing.

Since L^∞ is isomorphic to a $C(K)$, it follows that any map $L^\infty \to E$ is 2-summing. ∎

As a consequence of Theorems 17.1 and 17.2, if the spaces E and F are both of cotype 2, every Gauss-summing map is completely summing. (In Theorem 5.1 we dealt with the special case of Hilbert spaces.)

Grothendieck's Factorization Theorem (when $a = +\infty$)

We can now complete the discussion begun in Lecture 13. Suppose that we have $C(K) \overset{u}{\to} L^b(Y,\mu)$, $2 \geq b$. Because L^b is of cotype 2 (cf. p. 77), the map u factors as

$$C(K) \overset{(\alpha)}{\to} L^2(K,\lambda) \overset{v}{\to} L^b(Y,\mu)$$

with $\alpha \in L^2(K,\lambda)$. Next, because $L^2(K,\lambda)$ is of type 2, the map v factors as

$$L^2(K,\lambda) \overset{w}{\to} L^2(Y,\mu) \overset{(\beta)}{\to} L^b(Y,\mu)$$

where $\beta \in L^s$, $1/b = 1/s + 1/2$, as in our earlier discussion of the Grothendieck factorization theorem.

Thus we have the factorization

$$\begin{array}{ccc} C(K) & \xrightarrow{u} & L^b(Y,\mu) \\ {\scriptstyle(\alpha)}\downarrow & & \uparrow{\scriptstyle(\beta)} \\ L^2(K,\lambda) & \xrightarrow{w} & L^2(Y,\mu) \end{array}$$

with $\alpha \in L^2$ and $\beta \in L^s$, $1/b = 1/s + 1.2$.

Of course, L^∞ is a space $C(K)$, so we have a factorization for $a = +\infty$. But we emphasize that the measure λ is not fixed; it depends on u, in contrast to the case $a < +\infty$. ∎

The following is a result of Grothendieck characterizing Hilbert-Schmidt maps.

17.3. Theorem: Let H be a Hilbert space. For $u: H \to H$ to be Hilbert-Schmidt, the following conditions are each necessary and sufficient:

(a) u factors through L^∞.

(b) u factors through L^1.

Proof: If u is Hilbert-Schmidt, we know it is 2-summing, so it factors through L^∞: $H \to L^\infty \to L^2 \to H$. Now u Hilbert-Schmidt implies that u' is also Hilbert-Schmidt, so u' factors through L^∞ and therefore u factors through $(L^\infty)'$. But $(L^\infty)'$ = measures on some compact space, which is equivalent to some L^1 space. (If H is separable, we can factor through a separable L^1 space, namely that generated by the σ-field of the image of H in the big L^1 space.)

Conversely, any map $L^\infty \overset{u}{\to} H$ is 2-summing, by the factorization theorem:

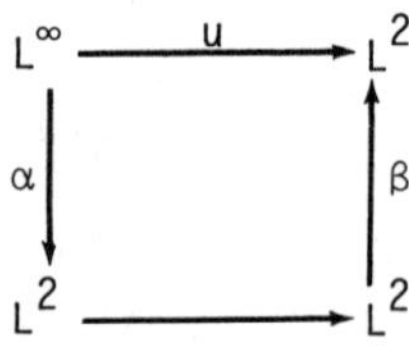

Hence u is Hilbert-Schmidt. ∎

Of course, Grothendieck's original proof of this result used none of our theory.

Finally, another result of Grothendieck:

17.4. Theorem: Every map $u: L^1 \to H$ is completely summing. (Grothendieck proved 1-summing; Maurey completed the result.)

Proof: The idea is to show that u is 2-summing. Then since L^1 has cotype 2, it follows that u is completely summing.

Suppose that $e = (e_n)$ is in $S\ell^2(L^1)$. Then it defines a map $\ell^2 \overset{e}{\to} L^1$. (Reason: If $e \in S\ell^2(E)$ it defines a map $E' \to \ell_2$. The transpose maps $\ell_2 \to E''$, but actually into E.) So we have the composite map

$$\ell^2 \overset{e}{\to} L^1 \overset{u}{\to} H \ .$$

Since this factors through L^1 it is Hilbert-Schmidt. Its Hilbert-Schmidt norm is

$$\{\sum_n |u(e_n)|^2\}^{1/2}$$

and this must be finite. Thus u is 2-summing. ∎

Let μ be a Radon measure on a Banach space E. We say that μ has *property TLC* ("central limit theorem" in French) provided that whenever $X_1,\ldots,X_n,\ldots$ are independent E-valued random variables, $\frac{1}{\sqrt{n}}\ (X_1+X_2+\cdots+X_n)$ converges in law to some limit Γ. Note: Γ must be a Gauss law.

If $E = \mathbb{R}$, this will be the case provided only that μ has mean 0 and finite second moment. More generally, μ must have scalar order 2. Thus we get a quadratic form associated to Γ and so, via reproducing kernels, some Hilbert space $H \to E$; Γ is then the image of the Gauss law on H.

Iterated Logarithm Law (ILL)

Consider $\frac{X_1+X_2+\cdots+X_n}{\sqrt{2n \log\log n}}$. First suppose $E = \mathbb{R}$. If the second moment of μ equals 1, this expression has a.s. lim sup = +1, lim inf = -1. More generally, when E is a Banach space we say that μ has *property ILL* if and only if the sequence

$$\frac{X_1+X_2+\cdots+X_n}{\sqrt{2n \log\log n}}$$

is relatively compact. (Then if μ has mean 0 and finite second moment, one can show that the set of accumulation points is precisely the image of the unit ball of some associated Hilbert space.)

Make the blanket assumption that $E(\mu) = 0$, i.e. barycenter is 0. Consider the following three properties:

(1) $\|\mu\|_2 < \infty$.

(2) μ has TLC .

(3) μ has ILL.

In general, no one of these implies another. But the first two imply the third. There are no other implications in general.

Relations with Type and Cotype: (1) For E a Banach space, the following are equivalent:

(a) $\|\mu\|_2 < +\infty \Rightarrow \mu$ has TLC

(b) E has type 2.

In this case $\|\mu\|_2 < +\infty$ also implies ILL.

(2) Suppose E has the property:

$$\|\mu\|_2 < +\infty \quad \Rightarrow \mu \text{ has ILL.}$$

This is equivalent to an inequality of the form

$$\{E|\sum_{i=1}^{n} \varepsilon_i x_i|^2\}^{1/2} \leq C\{\sum_{i=1}^{n} |x_i|^2\}^{1/2} \sqrt{\log\log n} \quad .$$

(The latter is implied by type 2, of course.)

Fact: This property $\Rightarrow$ E has type p for $p < 2$.

Proof: Suppose $\sum_i |x_i|^p < +\infty$. We want to show that $\sum \varepsilon_i x_i$ converges a.s. Arrange the terms so that $|x_1| \geq |x_2| \geq \cdots$. Then for $i \geq 1$ we have $|x_i|^p \leq C/i$, C some constant. So $|x_i|^2 \leq C(1/i)^{2/p}$. Hence

$$\left(\sum_m^n |x_i|^2\right)^{1/2} \leq C\{\sum_m^n (\tfrac{1}{i})^{2/p}\}^{1/2} \sim \frac{C}{m^{2/p-1/2}}$$

and this multiplied by $\sqrt{\log\log m}$ tends to 0 as $m, n \to \infty$. The result follows. ∎

Finally

(3) Suppose E is of cotype 2. Then

$$\mu \text{ TLC} \Rightarrow \|\mu\|_2 < +\infty \text{ and } \mu \text{ has ILL} \quad .$$

Lecture 18. Super-Reflexive Spaces. Modulus of Convexity, q-Convexity. "Trees" and Kelley-Chatterji Theorem. Enflo Theorem. Modulus of Smoothness, p-Smoothness. Properties Equivalent to Super-Reflexivity

Definition: A space E is *super-reflexive* provided that every F which is finitely representable in E is reflexive.

Examples: (1) Hilbert space

(2) L^p for $1 < p < \infty$ (by uniform convexity)

Reflexivity is *not* a superproperty. Here is an example of a reflexive space E such that L^∞ is finitely representable in E; so E is very far from super-reflexive. Let $(E_n)_{n\in\mathbb{N}}$ be a sequence of Banach spaces and form $E = \ell^2((E_n)_{n\in\mathbb{N}})$. Then $E' = \ell^2(E_n')_{n\in\mathbb{N}})$, so E is reflexive exactly when all the spaces E_n are reflexive. In particular $\ell^2((\ell^n)_{n\geq 2}) = E$ is reflexive. Obviously for any integer $p \geq 2$, $\ell^p \subseteq E$, so ℓ^∞ is finitely representable in E. (This space provides lots of counterexamples--it has practically every property that is not a superproperty.)

Recall that a space E is *uniformly convex* provided that for every $\varepsilon > 0$ there is a $\delta(\varepsilon) > 0$ such that if $|x|, |y| \leq 1$ and $|x-y| \geq \varepsilon$ then $|\frac{1}{2}(x+y)| \leq 1 - \delta(\varepsilon)$.

The *modulus of convexity* $\delta(\varepsilon)$ has been studied a great deal. Although $\delta(\varepsilon)$ is not a convex function of ε, one can define $\tilde{\delta}(\varepsilon)$ to be the greatest convex function $\leq \delta(\varepsilon)$. Then Figiel has shown that $\tilde{\delta}(\varepsilon) \geq \alpha\delta(\beta\varepsilon)$ for some constants $\alpha, \beta > 0$. So $\tilde{\delta}$ is "Orlicz equivalent" to δ. Accordingly we can take δ to be convex. It is trivial that $0 < \delta(\varepsilon) \leq \varepsilon$. But more is true:

18.1. Proposition: In any infinite dimensional uniformly convex Banach space,

$$\delta(\varepsilon) \lesssim \varepsilon^2 \quad (\text{meaning } \delta(\varepsilon) \leq \text{constant} \times \varepsilon^2) \quad .$$

Proof: To say that "δ is a modulus of convexity for E" is a superproperty, which really depends only on two-dimensional subspaces. Hence this property must hold for Hilbert space, which has modulus of convexity $\varepsilon^2/2$. Hence $\delta(\varepsilon) \leq \varepsilon^2/2$ for any infinite dimensional uniformly convex Banach space. ∎

Obviously, the closer $\delta(\varepsilon)$ is to ε^2, the greater is the uniform convexity. If $\delta(\varepsilon)$ decreases more rapidly, the unit ball is "flatter". Thus an inequality of the form $\delta(\varepsilon) \geq C\varepsilon^q$, $q \geq 2$, implies a reasonable amount of convexity.

18.2. Theorem (Figiel and Assouad): Equivalent to $\delta(\varepsilon) \geq C\varepsilon^q$ is the inequality $\frac{1}{2}(|x+y|^q + |x-y|^q) \geq |x|^q + L|y|^q$ for some constant L. (We call this property q-*convexity* of E.)

The proof, although "just a computation", is non-trivial. As a consequence we have the following.

18.3. Corollary: If ε^q is a modulus of convexity for E, then E has cotype q.

Proof: With $\varepsilon_i = \pm 1$, we have

$$\frac{1}{2}[|\varepsilon_1 x_1+\cdots+\varepsilon_{n-1}x_{n-1}+x_n|^q + |\varepsilon_1 x_1+\cdots+\varepsilon_{n-1}x_{n-1}-x_n|^q]$$
$$\geq |\varepsilon_1 x_1+\cdots+\varepsilon_{n-1}x_{n-1}|^q + \ell|x_n|^q .$$

Hence, by induction on n, we get

$$E_\varepsilon|\varepsilon_1 x_1+\cdots+\varepsilon_n x_n|^q \geq |x_1|^q + \ell(|x_2|^q+\cdots+|x_n|^q) .$$

This inequality shows that E is of cotype q. ∎

Remark: Recently Pisier proved that every uniformly convex Banach space can be *renormed* to be q-convex for some q.

Examples: The modulus of convexity of L^q is ε^r, where $r = \max(q,2)$, except that $\delta \equiv 0$ for L^1 and L^∞ since these are not uniformly convex.

It is well known that a uniformly convex space E is reflexive. It follows immediately that actually E must be super-reflexive. Now reflexity is purely a *topological* property; it does not depend on the norm. Therefore the same is true of super-reflexivity. Accordingly a *uniformly convexifiable* space (one with an equivalent uniformly convex norm) is super-reflexive.

<u>Remark</u>: The general situation about superproperties is the following. Let P be a given superproperty. Let Q be the property: "E is isomorphic to a Banach space E_1 which satisfies P", or equivalently, "E can be renormed so as to satisfy P". Then Q is a superproperty too. Indeed, suppose that F is finitely representable in E; then it is a subspace of an ultrapower $E^I/\mathcal{U}$. But since E is isomorphic to E_1, $E^I/\mathcal{U}$ is isomorphic to $E_1^I/\mathcal{U}$, whence F is isomorphic to a subspace F_1 of $E_1^I/\mathcal{U}$; F_1 satisfies P, so F satisfies Q.

Every reflexive space E has the *Radon-Nikodym Property* (RNP): Every E-valued measure absolutely continuous with respect to a positive measure is an integral. (<u>Note</u>: The most important other class of spaces with RNP are the separable dual spaces, such as ℓ^1.)

Then if E is super-reflexive, E has the super-RNP.

Related to all this is a theorem of J. L. Kelley. We need the notion of a *tree* in a Banach space E. One constructs it as follows. Start with a point x_0, the "0-branch". Then construct the "1-branches" x_1, x_{-1} such that $x_0 = \frac{1}{2}(x_1+x_{-1})$. Next construct four "2-branches" $x_{\pm 1,\pm 1}$ so that $x_{\varepsilon_1} = \frac{1}{2}[x_{\varepsilon_1,1} + x_{\varepsilon_1,-1}]$ for $\varepsilon_1 = \pm 1$. In general, at the kth level there are 2^k k-branches, and we have

$$x_{\varepsilon_1,\ldots,\varepsilon_k} = \frac{1}{2}[x_{\varepsilon_1,\ldots,\varepsilon_k,1} + x_{\varepsilon_1,\ldots,\varepsilon_k,-1}] \quad .$$

Picture:

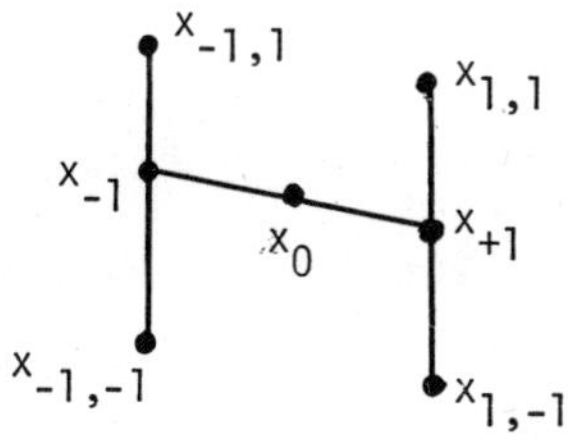

We refer to a tree with branches through level n as an n-*tree*. (Such a tree will have $2^{n+1}-1$ "branches".) Also, all the branches are supposed to lie in the unit ball. We have an (n,ε)-tree if it is an n-tree with all branches of length $\geq \varepsilon$.

We say that E *has infinite trees* provided that there exists $\varepsilon > 0$ such that E has (∞,ε)-trees in the unit ball.

E *has no infinite trees* provided that, for all $\varepsilon > 0$, there is no (∞,ε)-tree in the unit ball.

We say E *has large finite trees* provided that there exists $\varepsilon > 0$ such that E has arbitrarily large ε-trees in the unit ball.

Finally, E *has no large finite trees* provided that, for every $\varepsilon > 0$, there is an integer $N(\varepsilon)$ so that every ε-tree in the unit ball has at most $N(\varepsilon)$ branches.

The latter property is a superproperty. In fact, Super(E has no infinite trees) = (E has no large finite trees). The proof is omitted.

18.4. Theorem (Kelley, improved by Chatterji): If E is reflexive, E has no infinite trees. More generally this is so if E has the RNP.

Chatterji's proof uses martingales. (Clearly a tree "is" a martingale.) The basic idea is to show that the RNP is equivalent to the property that every bounded martingale is a.s. convergent. But the bounded martingale corresponding to an infinite tree is *nowhere* convergent since the branches have lengths $\geq \varepsilon > 0$.

Suppose now that E has the *super*-RNP. Then, by the Kelley-Chatterji theorem, E has the super version of "no infinite trees"; that is, E has no large finite trees.

Thus we have the implications:

uniformly convexifiable $\Rightarrow$ super-reflexive $\Rightarrow$ super-RNP
$\Rightarrow$ no large finite trees.

Enflo closed the circle by proving the following great result.

18.5. Theorem (Enflo): E has no large finite trees $\Rightarrow$ E is uniformly convexifiable.

Accordingly we have the four equivalences:

uniformly convexifiable $\Leftrightarrow$ super-reflexive $\Leftrightarrow$ super-RNP
$\Leftrightarrow$ no large finite trees.

Note that in this indirect way we have proved also that uniform convexifiability is a superproperty. (Is there a more direct proof?)

Uniformly Smooth Spaces

Let E be a Banach space, $x, y \in E$, with $|x|, |y| \leq 1$. Consider the expression

$$\frac{1}{2}[|x+\tau y| + |x-\tau y|] - |x| \quad .$$

Obviously this is always $\leq |\tau|$. Call the supremum $\rho(\tau)$, the *modulus of smoothness*. We say that E is *uniformly smooth* provided that $\rho(\tau) = o(\tau)$ as $\tau \downarrow 0$. Uniform smoothness means that the norm $|\cdot|$ is Frechet differentiable away from 0, and it is uniformly so on the surface of the unit sphere.

Examples:

(a) (b)

(a) represents a uniformly convex norm which is not uniformly smooth; (b) represents the reverse.

We always have $\rho(\tau) \geq c\tau^2$ for an infinite dimensional Banach space; τ^2 holds in Hilbert spaces. We say that E is *p-smooth*, $1 < p \leq 2$, provided that $\rho(\tau) \leq c\tau^p$. This is the property "dual" to the notion of q-convexity mentioned earlier, and the next theorem is the analogue of 18.2.

18.6. Theorem (Pisier-Assouad): E is p-smooth $\Leftrightarrow$ for some constant L the following inequality holds:

$$\frac{1}{2}[|x+y|^p + |x-y|^p] \leq |x|^p + L|y|^p \quad .$$

It is also a fact that every uniformly smooth space is reflexive--indeed super-reflexive.

Corresponding to uniform convexifiability, there is a super-property of *uniform smoothability*: the existence of an equivalent uniformly smooth norm. (One can then find arbitrarily close norms that are uniformly smooth.)

The theorem of Pisier which we mentioned following 18.3 says that if E is uniformly convex, then it is q-convex, after renorming, for some q. A sharper result can be stated: suppose that $\delta(\varepsilon)/\varepsilon^q \to \infty$ as $\varepsilon \to 0$. Then for some $q_1 < q$, E is q_1-convexifiable.

This theorem of Pisier has an analogue for smoothness. In particular, if E is super-reflexive then it can be renormed to be p-smooth for some p, $1 < p \leq 2$. Actually in this case one can make E simultaneously p-smooth and q-convex by arbitrarily small perturbations of its original norm: $\|\cdot\|_1 + \varepsilon\|\cdot\|_2 + \varepsilon\|\cdot\|_3$, where $\|\cdot\|_2$ and $\|\cdot\|_3$ are respectively p_0-smooth and q_0-convex, is both p-smooth and q-convex. But the indices p, q depend on E.

The explanation is that E is p-smooth $\Leftrightarrow$ E' is p'-convex, and conversely. Thus smoothness and convexity are dual properties.

Here it is useful to recall Young's duality for convex functions. Let $\phi(x)$ be a (convex) function on $\mathbb{R}$, and define

$$\phi^*(\xi) = \sup_x (x\xi - \phi(x)) \quad .$$

$\phi^*(\xi)$ is always convex, whether or not ϕ is. Then $\phi^{**}(x)$ is the largest convex minorant of ϕ.

Although the modulus of convexity $\delta(\varepsilon)$ is not convex, the modulus of smoothness $\rho(\tau)$ is convex by its very definition. Moreover, if E' is dual to E,

$$\rho_{E'}(\tau) = \sup_\varepsilon \left[\frac{\tau\varepsilon}{2} - \delta_E(\varepsilon)\right] \quad .$$

That is, $\rho_{E'}(\tau) = \delta_E^*(\tau/2)$, so that $\rho_{E'}$ is (almost) the Young dual of δ_E. The second dual is the largest convex minorant of δ_E.

Equivalent are: $\rho_{E'}(\tau) = o(\tau)$; $\delta_E(\varepsilon) > 0$. Thus E is uniformly smooth $\Leftrightarrow$ E' is uniformly convex. More precisely, E is p-smooth $\Leftrightarrow$ E' is p'-convex. Therefore every theorem for convexity has its "dual" for smoothness.

Note: The "dual" of a super-property is not necessarily a super-property. However, using convexity and smoothness, one sees that the dual of a super-reflexive space is super-reflexive.

Exactly as for uniformly convex spaces, there is a direct proof that a uniformly smooth Banach space is reflexive. And hence a uniformly smooth space is super-reflexive. Accordingly, since super-reflexivity is a topological property, it follows that a *uniformly smoothable* space is super-reflexive.

To summarize, we have the following result.

18.7. Theorem: The following properties of a Banach space E are equivalent:

(1) E is super-reflexive.

(2) E has the super-RNP.

(3) E is uniformly convexifiable.

(4) E can be renormed so that for some $q \geq 2$, the modulus of convexity $\delta(\varepsilon) \geq c\varepsilon^q$.

(5) E is uniformly smoothable.

(6) E can be renormed so that for some $p \leq 2$, the modulus of smoothness $\rho(\tau) \leq c\tau^p$.

(7) E has no large trees.

(8) E' has one (hence all) of the properties above.

The argument goes as follows: E uniformly smoothable $\Rightarrow$ E super-reflexive $\Rightarrow$ E uniformly convexifiable $\Rightarrow$ E' uniformly smoothable $\Rightarrow$ E' super-reflexive $\Rightarrow$ E' uniformly convexifiable $\Rightarrow$ E uniformly smoothable. Hence all the listed properties are equivalent, and in particular uniform smoothability is a super-property.

Remarks: 1. A super-reflexive space E must have type Gauss 1, since L^1, being non-reflexive, cannot be finitely representable in E. But type Gauss 1 implies 1-Pietsch. Hence a super-reflexive space E is of type $(1+\varepsilon)$-Gauss for some $\varepsilon > 0$; but nothing more can be specified. Thus, for $1 < p < p_0$, the space L^p is super-reflexive but not of type p_0-Gauss. In the converse direction there are counterexamples: if $p < 2$ then James produced spaces of type p-Gauss which are *not* super-reflexive. Recently, he even constructed an example with p=2.

2. We say that a space E is *ergodic* provided that, for every linear contraction u on E and every x in E, the sequence

$$\frac{1}{n}(x + ux + u^2x + \cdots + u^{n-1}x)$$

converges to a fixed point of u.

If the trajectory of every x lies in a weakly compact subset of E, then it is known that u is ergodic in that the above limit exists. So in particular if E is reflexive it is ergodic (but the converse is false). Hence super-reflexive ⇒ super-ergodic. Here Brunel and Sucheston have proved the converse: super-ergodic ⇒ super-reflexive. Hence to the list of equivalent properties in 18.7 we can add:

(9) E is super-ergodic.

3. Definition: A sequence is a *strong Banach-Saks sequence* provided that all its subsequences are Cesaro-summable to the same limit, uniformly. E is a *Banach-Saks space* provided that each bounded sequence in E has a strongly Banach-Saks subsequence.

This implies that E is reflexive. Another property equivalent to super-reflexivity is:

(10) E is super-Banach-Saks.

Lecture 19. Martingale Type and Cotype. Results of Pisier. Twelve Properties Equivalent to Super-reflexivity. Type for Subspaces of L^p (Rosenthal Theorem)

We now turn to some applications of *martingales*. Let $(X_n)_{n\in\mathbb{N}}$ be a family of real valued random variables associated to σ-fields $\mathcal{T}_n$, and suppose that this is a martingale, so that X_n is the conditional expectation of X_{n+1} with respect to $\mathcal{T}_n$.

Define $X_* = \sup_n X_n$. Also define the "increments" by $dX_0 = X_0$,

$dX_n = X_n - X_{n-1}$ for $n \geq 1$. Set $[X,X]^{1/2} = (\sum_n dX_n{}^2)^{1/2}$.

The *Burkholder-Davis-Gundy (BDG) inequality* says that the L^p norms of X_* and $[X,X]^{1/2}$ are equivalent. That is, for $1 \leq a \leq A < +\infty$, there are universal constants c, C such that

$$c(E[X,X]^{a/2})^{1/a} \leq (E\, X_*{}^a)^{1/a} \leq C(E[X,X]^{a/2})^{1/a} \quad .$$

As a special case, consider the following martingale. Let $(x_i)_{i\geq 0}$ be a sequence of real numbers and set $X_n = \varepsilon_0 x_0 + \varepsilon_1 x_1 + \cdots + \varepsilon_n x_n$, $n = 0,1,2,\ldots$. Then the BDG inequality reduces to *Khintchine's inequality*:

$$E\left(|\sum \varepsilon_k x_k|^a\right)^{1/a} \sim (E|X_*|^a)^{1/a} \approx (\sum|x_n|^2)^{1/2} \quad .$$

So we can think of the BDG inequality as a generalization of Khintchine's inequality. Now, just as Khintchine's inequality suggested the notion of type Rademacher-p, the BDG inequality suggests the following notion.

<u>Definition</u>: Let $1 \leq p \leq 2$. A Banach space E is of *type* p-*martingale* provided that, for $1 \leq a < +\infty$, there is a constant $C = C_a$ such that every martingale (X_n) in E satisfies the inequality

$$(*) \qquad (E|X_*|^a)^{1/a} \leq C_a\left(E(\sum_n |dX_n|^p)^{a/p}\right)^{1/a} \quad .$$

<u>Note</u>: By an observation of Assouad and Pisier, we get the same property whether we assume (*) for *some* a in $[1,\infty)$ or for *all* such a.

This property of course implies type p-Gauss, p-Rademacher, etc.; these have to do with particular martingales.

Similarly, we say that E is of *cotype* q-*martingale*, $2 \leq q < +\infty$, provided that for some (equivalently, every) a, $1 \leq a < +\infty$, we have

$$[E(X^a)]^{1/a} \geq c\left(E(\sum_n |dX_n|^q)^{a/q}\right)^{1/a} \quad .$$

19.1. <u>Theorem</u> (Assouad-Pisier): E is q-convex $\Leftrightarrow$ E has the following "metric" version of the cotype q-martingale property:

$$E|X_*|^q \geq E|X_0|^q + L\cdot E(\sum_n |dX_n|^q)$$

where L is some constant.

(There is an analogous equivalence between p-smoothness and "metric" type p-martingale. Cf. 18.2 and 18.6.)

Note that q-convexity depends on the norm of E, not merely the topology. So, to be consistent, we must have a version of cotype q-martingale which depends on the norm, reflected in the coefficient 1 of $E|X_0|^q$.

A related, more general result is the following.

19.2. <u>Theorem</u> (Pisier): (1) A Banach space E is of cotype q-martingale $\Leftrightarrow$ E is q-convexifiable, i.e. has an equivalent q-convex norm.

(2) E is of type p-martingale $\Leftrightarrow$ E is p-smoothable, i.e. has an equivalent p-smooth norm.

19.3. <u>Theorem</u> (Pisier): If E is uniformly convexifiable, it is of cotype q-martingale for some q. Hence E has an equivalent q-convex norm, for some q. (The corresponding result for uniformly smoothable spaces is also true; every such space is p-smoothable for some p.)

Some results related to 19.3 were proved by R. C. James and A. Beck; the martingale version is Pisier's.

19.4. <u>Proposition</u>: If E is of cotype q-martingale, then E has no large trees.

<u>Proof</u>: Indeed, for the martingale associated with a tree (cf. Lecture 18), $X_* \leq 1$, since the tree lies in the unit ball of E. Hence

$$1 \geq E(X^a)^{1/a} \geq c\cdot\varepsilon\cdot n^{1/q}$$

because the branches of an (n,ε)-tree are at least ε in length. Hence n cannot be very large; in fact, $n \leq \text{constant}/\varepsilon^q$. ∎

Thus we have the following chain of implications: E super-reflexive ⇒ E uniformly convexifiable ⇒ E q-convexifiable for some q ⇒ E of cotype q-martingale ⇒ E has no large trees ⇒ E is super-reflexive.

This also shows that "E of cotype q-martingale for some q" is a superproperty.

Similarly, the following implications hold: E type p-martingale for some p ⇒ E' cotype p'-martingale where p' is the conjugate index (a direct elementary proof is possible) ⇒ E' is super-reflexive ⇒ E' is q-convex for some q ⇒ E is q'-smooth ⇒ E is type p-martingale for some p(= q').

To summarize:

19.5. Theorem: The following properties of a Banach space E are equivalent:

(1) E is super-reflexive.

(2) E has the super-RNP.

(3) E is uniformly convexifiable.

(4) E is of cotype q-martingale for some q.

(5) For some q, E has an equivalent norm with $\delta(\varepsilon) \geq c\varepsilon^q$.

(6) E is uniformly smoothable.

(7) E is of type p-martingale for some p.

(8) For some p, E has an equivalent norm with $\rho(\tau) \leq c\tau^p$.

(9) E has no large trees.

(10) E is super-ergodic.

(11) E is super-Banach-Saks.

(12) E' has one (hence all) of the properties listed above.

Remarks: In accordance with the observation we made following 18.3, we can state the following. Let p and q be *fixed*. Since p-smoothness and q-convexity are superproperties, so are p-smoothability and q-convexifiability. And Pisier proved precisely that p-smoothability = type p-martingale, and q-convexifiability = cotype q-martingale. Thus type p-martingale and cotype q-martingale (for *fixed* p and q) are superproperties; this can also be seen directly from their definitions.

We will conclude with a theorem of H. Rosenthal ("On subspaces of L^p", Ann. of Math. (ser. 2) 97 (1973), 344-373.)

19.6. Theorem (Rosenthal): Let E be a subspace of L^p, $1 \leq p < 2$. Then the following properties are equivalent:

(1) The topology of E is induced by L^0 (equivalently, by some L^r for $r < p$).

(2) E contains no subspace isomorphic to ℓ^p.

(2') E contains no complemented subspace isomorphic to ℓ^p.

(2") L^p is not finitely representable in E.

(3) $E \subseteq$ some L^q, $q > p$.

(4) E has type Gauss-p (which L^p does not have).

(5) Every linear map $u: E \to \ell^p$ is compact.

(6) (If $p = 1$) E is reflexive.

Proof: The implications (2") ⇒ (2) ⇒ (2') are easy or trivial. That (2') ⇒ (1) is a theorem of Kadeč and Pelczynski.

(1) ⇒ (4) is easy. For consider $E \hookrightarrow L^p \hookrightarrow L^r$ for $r < p$. This map $E \to L^r$ factors through L^p and therefore (by Maurey's Theorem) is of type Gauss-p. But this is the *identity* $E \to E$ since the topology of E is the L^r topology by (1). Hence E is of type Gauss-p.

(4) ⇒ (3): $E \hookrightarrow L^p$. Since E is type Gauss-p and the type interval is open, E is of type Gauss-p_1 for some $p_1 > p$. So the map $E \hookrightarrow L^p$ factors through some L^q, $q > p$, i.e. $E \subset L^q$ for some $q > p$.

(4) ⇒ (2") is clear, since L^p does not have type Gauss-p.

(3) ⇒ (4) is easy.

Thus we have shown that (1)-(4) are equivalent.

Now consider (5). If every map $E \to \ell^p$ is compact, it follows that ℓ^p cannot be a complemented subspace of E --else ℓ^p would be the range of a (non-compact) projection. Hence (5) ⇒ (2').

Conversely, (4) ⇒ (5): $E \xrightarrow{u} \ell^p$ factors through ℓ^q, $q > p$:

$$\begin{array}{ccc} E & \longrightarrow & \ell^p \\ & \searrow \quad \nearrow h = (\alpha) & \\ & \ell^q & \end{array}$$

So (since h is compact) the map u is compact.

Thus (5) is equivalent to the earlier properties.

Finally, when $p = 1$, (6) is equivalent to all the other properties. For if E is reflexive, it cannot contain L^1; on the other hand, all the other properties imply $E \subseteq L^{1+\varepsilon}$, so E is reflexive. ∎

Thus we have given a rather short proof of Rosenthal's Theorem. Of course, at the time, our machinery was not available to Rosenthal. For example, the openness of the type interval was not known. Rosenthal proved the openness of type interval for subspaces of L^p, and it was this that suggested that it might be valid in general.

Stop press:

Very recently (August, 1980) Pisier established the following interesting result , which had been an outstanding conjecture.

With each Banach space E there are associated numbers $p(E)$ and $q(E)$ defined as follows:

(1) $p(E) - \sup\{p : E \text{ has type } p\}$

$= \min\{p : L^p \text{ is finitely representable in } E\}$

where the equality follows from Theorem 12.9

(2) $q(E) = \inf\{q : E \text{ has cotype } q\}$

$= \max\{q : L^q \text{ is finitely representable in } E\}$

by Theorems 16.2, 16.5.

Then, by Theorem 16.2, the Hölder conjugate of $q(E)$ is given by

(3) $q(E)' = \sup\{p : E \text{ is } p\text{-Pietsch}\}$.

Now, as the L^r spaces show, surely $q(E)$ and $p(E)$ are not conjugate. However, by Theorem 16.6, we have

(4) $q(E') \leq p(E)'$.

Moreover, in many cases (4) is an equality; e.g. for $E = L^r$ with $1 < r < \infty$. Pisier's theorem is much more general:

Theorem: If E has Gauss type 1 (or equivalently, by Ths. 12.7 and 15.4, if E' has, or equivalently $p(E) > 1$ or $p(E') > 1$) then $q(E') = p(E)'$ and $q(E) = p(E')'$.

REFERENCES

1. Séminaire L. Schwartz 1969-70, "Applications Radonifiantes", École Polytechnique, Paris.
2. Séminaire Maurey-Schwartz 1972-73, "Espaces L^p et Applications Radonifiantes", École Polytechnique, Paris.
3. Séminaire Maurey-Schwartz 1973-74, "Espaces L^p, Applications Radonifiantes, et Geometrie des Espaces de Banch", École Polytechnique, Paris.
4. Seminar Schwartz, Part I, "Cylindrical Probabilities and p-summing and p-Radonifying Maps", Notes on Pure Mathematics No. 7, 1973, Australian National University, Canberra.
5. G. Pisier, Martingales with values in uniformly convex spaces, Israel Jour. of Math. 20 (1975), 326-350.

The first four are general references. The details of the theory of p-summing maps, cylindrical probabilities, etc. will be found in [1] and [4] along with citations of the journal literature. Many of the seminar talks in [2] and [3] are also relevant. Explicit mention should be made of those by Assouad, Beauzamy, Maurey, and Pisier. Finally, Pisier's paper [5] contains the proofs of some of the main results discussed in Lectures 18 and 19, along with useful bibliographic references to the work of Enflo, James, and others.

INDEX

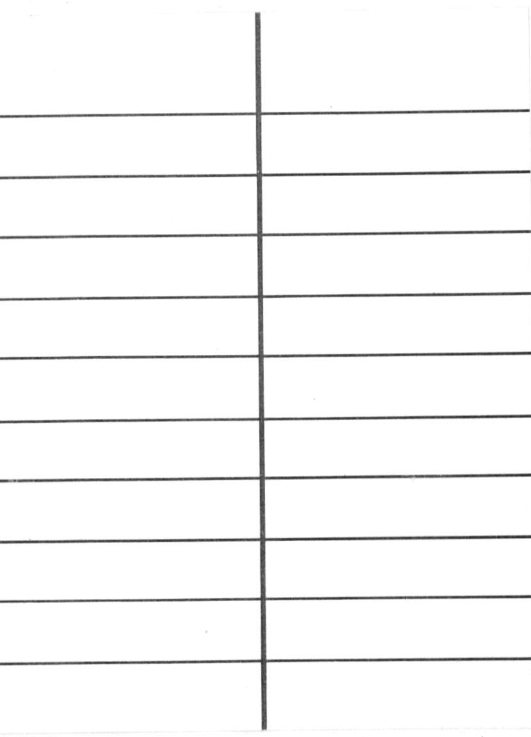

Vol. 700: Module Theory, Proceedings, 1977. Edited by C. Faith and S. Wiegand. X, 239 pages. 1979.

Vol. 701: Functional Analysis Methods in Numerical Analysis, Proceedings, 1977. Edited by M. Zuhair Nashed. VII, 333 pages. 1979.

Vol. 702: Yuri N. Bibikov, Local Theory of Nonlinear Analytic Ordinary Differential Equations. IX, 147 pages. 1979.

Vol. 703: Equadiff IV, Proceedings, 1977. Edited by J. Fábera. XIX, 441 pages. 1979.

Vol. 704: Computing Methods in Applied Sciences and Engineering, 1977, I. Proceedings, 1977. Edited by R. Glowinski and J. L. Lions. VI, 391 pages. 1979.

Vol. 705: O. Forster und K. Knorr, Konstruktion verseller Familien kompakter komplexer Räume. VII, 141 Seiten. 1979.

Vol. 706: Probability Measures on Groups, Proceedings, 1978. Edited by H. Heyer. XIII, 348 pages. 1979.

Vol. 707: R. Zielke, Discontinuous Čebyšev Systems. VI, 111 pages. 1979.

Vol. 708: J. P. Jouanolou, Equations de Pfaff algébriques. V, 255 pages. 1979.

Vol. 709: Probability in Banach Spaces II. Proceedings, 1978. Edited by A. Beck. V, 205 pages. 1979.

Vol. 710: Séminaire Bourbaki vol. 1977/78, Exposés 507–524. IV, 328 pages. 1979.

Vol. 711: Asymptotic Analysis. Edited by F. Verhulst. V, 240 pages. 1979.

Vol. 712: Equations Différentielles et Systèmes de Pfaff dans le Champ Complexe. Edité par R. Gérard et J.-P. Ramis. V, 364 pages. 1979.

Vol. 713: Séminaire de Théorie du Potentiel, Paris No. 4. Edité par F. Hirsch et G. Mokobodzki. VII, 281 pages. 1979.

Vol. 714: J. Jacod, Calcul Stochastique et Problèmes de Martingales. X, 539 pages. 1979.

Vol. 715: Inder Bir S. Passi, Group Rings and Their Augmentation Ideals. VI, 137 pages. 1979.

Vol. 716: M. A. Scheunert, The Theory of Lie Superalgebras. X, 271 pages. 1979.

Vol. 717: Grosser, Bidualräume und Vervollständigungen von Banachmoduln. III, 209 pages. 1979.

Vol. 718: J. Ferrante and C. W. Rackoff, The Computational Complexity of Logical Theories. X, 243 pages. 1979.

Vol. 719: Categorial Topology, Proceedings, 1978. Edited by H. Herrlich and G. Preuß. XII, 420 pages. 1979.

Vol. 720: E. Dubinsky, The Structure of Nuclear Fréchet Spaces. V, 187 pages. 1979.

Vol. 721: Séminaire de Probabilités XIII. Proceedings, Strasbourg, 1977/78. Edité par C. Dellacherie, P. A. Meyer et M. Weil. VII, 647 pages. 1979.

Vol. 722: Topology of Low-Dimensional Manifolds. Proceedings, 1977. Edited by R. Fenn. VI, 154 pages. 1979.

Vol. 723: W. Brandal, Commutative Rings whose Finitely Generated Modules Decompose. II, 116 pages. 1979.

Vol. 724: D. Griffeath, Additive and Cancellative Interacting Particle Systems. V, 108 pages. 1979.

Vol. 725: Algèbres d'Opérateurs. Proceedings, 1978. Edité par P. de la Harpe. VII, 309 pages. 1979.

Vol. 726: Y.-C. Wong, Schwartz Spaces, Nuclear Spaces and Tensor Products. VI, 418 pages. 1979.

Vol. 727: Y. Saito, Spectral Representations for Schrödinger Operators With Long-Range Potentials. V, 149 pages. 1979.

Vol. 728: Non-Commutative Harmonic Analysis. Proceedings, 1978. Edited by J. Carmona and M. Vergne. V, 244 pages. 1979.

Vol. 729: Ergodic Theory. Proceedings, 1978. Edited by M. Denker and K. Jacobs. XII, 209 pages. 1979.

Vol. 730: Functional Differential Equations and Approximation of Fixed Points. Proceedings, 1978. Edited by H.-O. Peitgen and H.-O. Walther. XV, 503 pages. 1979.

Vol. 731: Y. Nakagami and M. Takesaki, Duality for Crossed Products of von Neumann Algebras. IX, 139 pages. 1979.

Vol. 732: Algebraic Geometry. Proceedings, 1978. Edited by K. Lønsted. IV, 658 pages. 1979.

Vol. 733: F. Bloom, Modern Differential Geometric Techniques in the Theory of Continuous Distributions of Dislocations. XII, 206 pages. 1979.

Vol. 734: Ring Theory, Waterloo, 1978. Proceedings, 1978. Edited by D. Handelman and J. Lawrence. XI, 352 pages. 1979.

Vol. 735: B. Aupetit, Propriétés Spectrales des Algèbres de Banach. XII, 192 pages. 1979.

Vol. 736: E. Behrends, M-Structure and the Banach-Stone Theorem. X, 217 pages. 1979.

Vol. 737: Volterra Equations. Proceedings 1978. Edited by S.-O. Londen and O. J. Staffans. VIII, 314 pages. 1979.

Vol. 738: P. E. Conner, Differentiable Periodic Maps. 2nd edition, IV, 181 pages. 1979.

Vol. 739: Analyse Harmonique sur les Groupes de Lie II. Proceedings, 1976–78. Edited by P. Eymard et al. VI, 646 pages. 1979.

Vol. 740: Séminaire d'Algèbre Paul Dubreil. Proceedings, 1977–78. Edited by M.-P. Malliavin. V, 456 pages. 1979.

Vol. 741: Algebraic Topology, Waterloo 1978. Proceedings. Edited by P. Hoffman and V. Snaith. XI, 655 pages. 1979.

Vol. 742: K. Clancey, Seminormal Operators. VII, 125 pages. 1979.

Vol. 743: Romanian-Finnish Seminar on Complex Analysis. Proceedings, 1976. Edited by C. Andreian Cazacu et al. XVI, 713 pages. 1979.

Vol. 744: I. Reiner and K. W. Roggenkamp, Integral Representations. VIII, 275 pages. 1979.

Vol. 745: D. K. Haley, Equational Compactness in Rings. III, 167 pages. 1979.

Vol. 746: P. Hoffman, τ-Rings and Wreath Product Representations. V, 148 pages. 1979.

Vol. 747: Complex Analysis, Joensuu 1978. Proceedings, 1978. Edited by I. Laine, O. Lehto and T. Sorvali. XV, 450 pages. 1979.

Vol. 748: Combinatorial Mathematics VI. Proceedings, 1978. Edited by A. F. Horadam and W. D. Wallis. IX, 206 pages. 1979.

Vol. 749: V. Girault and P.-A. Raviart, Finite Element Approximation of the Navier-Stokes Equations. VII, 200 pages. 1979.

Vol. 750: J. C. Jantzen, Moduln mit einem höchsten Gewicht. III, 195 Seiten. 1979.

Vol. 751: Number Theory, Carbondale 1979. Proceedings. Edited by M. B. Nathanson. V, 342 pages. 1979.

Vol. 752: M. Barr, *-Autonomous Categories. VI, 140 pages. 1979.

Vol. 753: Applications of Sheaves. Proceedings, 1977. Edited by M. Fourman, C. Mulvey and D. Scott. XIV, 779 pages. 1979.

Vol. 754: O. A. Laudal, Formal Moduli of Algebraic Structures. III, 161 pages. 1979.

Vol. 755: Global Analysis. Proceedings, 1978. Edited by M. Grmela and J. E. Marsden. VII, 377 pages. 1979.

Vol. 756: H. O. Cordes, Elliptic Pseudo-Differential Operators – An Abstract Theory. IX, 331 pages. 1979.

Vol. 757: Smoothing Techniques for Curve Estimation. Proceedings, 1979. Edited by Th. Gasser and M. Rosenblatt. V, 245 pages. 1979.

Vol. 758: C. Năstăsescu and F. Van Oystaeyen; Graded and Filtered Rings and Modules. X, 148 pages. 1979.

Vol. 759: R. L. Epstein, Degrees of Unsolvability: Structure and Theory. XIV, 216 pages. 1979.

Vol. 760: H.-O. Georgii, Canonical Gibbs Measures. VIII, 190 pages. 1979.

Vol. 761: K. Johannson, Homotopy Equivalences of 3-Manifolds with Boundaries. 2, 303 pages. 1979.

Vol. 762: D. H. Sattinger, Group Theoretic Methods in Bifurcation Theory. V, 241 pages. 1979.

Vol. 763: Algebraic Topology, Aarhus 1978. Proceedings, 1978. Edited by J. L. Dupont and H. Madsen. VI, 695 pages. 1979.

Vol. 764: B. Srinivasan, Representations of Finite Chevalley Groups. XI, 177 pages. 1979.

Vol. 765: Padé Approximation and its Applications. Proceedings, 1979. Edited by L. Wuytack. VI, 392 pages. 1979.

Vol. 766: T. tom Dieck, Transformation Groups and Representation Theory. VIII, 309 pages. 1979.

Vol. 767: M. Namba, Families of Meromorphic Functions on Compact Riemann Surfaces. XII, 284 pages. 1979.

Vol. 768: R. S. Doran and J. Wichmann, Approximate Identities and Factorization in Banach Modules. X, 305 pages. 1979.

Vol. 769: J. Flum, M. Ziegler, Topological Model Theory. X, 151 pages. 1980.

Vol. 770: Séminaire Bourbaki vol. 1978/79 Exposés 525–542. IV, 341 pages. 1980.

Vol. 771: Approximation Methods for Navier-Stokes Problems. Proceedings, 1979. Edited by R. Rautmann. XVI, 581 pages. 1980.

Vol. 772: J. P. Levine, Algebraic Structure of Knot Modules. XI, 104 pages. 1980.

Vol. 773: Numerical Analysis. Proceedings, 1979. Edited by G. A. Watson. X, 184 pages. 1980.

Vol. 774: R. Azencott, Y. Guivarc'h, R. F. Gundy, Ecole d'Eté de Probabilités de Saint-Flour VIII-1978. Edited by P. L. Hennequin. XIII, 334 pages. 1980.

Vol. 775: Geometric Methods in Mathematical Physics. Proceedings, 1979. Edited by G. Kaiser and J. E. Marsden. VII, 257 pages. 1980.

Vol. 776: B. Gross, Arithmetic on Elliptic Curves with Complex Multiplication. V, 95 pages. 1980.

Vol. 777: Séminaire sur les Singularités des Surfaces. Proceedings, 1976-1977. Edited by M. Demazure, H. Pinkham and B. Teissier. IX, 339 pages. 1980.

Vol. 778: SK1 von Schiefkörpern. Proceedings, 1976. Edited by P. Draxl and M. Kneser. II, 124 pages. 1980.

Vol. 779: Euclidean Harmonic Analysis. Proceedings, 1979. Edited by J. J. Benedetto. III, 177 pages. 1980.

Vol. 780: L. Schwartz, Semi-Martingales sur des Variétés, et Martingales Conformes sur des Variétés Analytiques Complexes. XV, 132 pages. 1980.

Vol. 781: Harmonic Analysis Iraklion 1978. Proceedings 1978. Edited by N. Petridis, S. K. Pichorides and N. Varopoulos. V, 213 pages. 1980.

Vol. 782: Bifurcation and Nonlinear Eigenvalue Problems. Proceedings, 1978. Edited by C. Bardos, J. M. Lasry and M. Schatzman. VIII, 296 pages. 1980.

Vol. 783: A. Dinghas, Wertverteilung meromorpher Funktionen in ein- und mehrfach zusammenhängenden Gebieten. Edited by R. Nevanlinna and C. Andreian Cazacu. XIII, 145 pages. 1980.

Vol. 784: Séminaire de Probabilités XIV. Proceedings, 1978/79. Edited by J. Azéma and M. Yor. VIII, 546 pages. 1980.

Vol. 785: W. M. Schmidt, Diophantine Approximation. X, 299 pages. 1980.

Vol. 786: I. J. Maddox, Infinite Matrices of Operators. V, 122 pages. 1980.

Vol. 787: Potential Theory, Copenhagen 1979. Proceedings, 1979. Edited by C. Berg, G. Forst and B. Fuglede. VIII, 319 pages. 1980.

Vol. 788: Topology Symposium, Siegen 1979. Proceedings, 1979. Edited by U. Koschorke and W. D. Neumann. VIII, 495 pages. 1980.

Vol. 789: J. E. Humphreys, Arithmetic Groups. VII, 158 pages. 1980.

Vol. 790: W. Dicks, Groups, Trees and Projective Modules. IX, 127 pages. 1980.

Vol. 791: K. W. Bauer and S. Ruscheweyh, Differential Operators for Partial Differential Equations and Function Theoretic Applications. V, 258 pages. 1980.

Vol. 792: Geometry and Differential Geometry. Proceedings, 1979. Edited by R. Artzy and I. Vaisman. VI, 443 pages. 1980.

Vol. 793: J. Renault, A Groupoid Approach to C*-Algebras. III, 160 pages. 1980.

Vol. 794: Measure Theory, Oberwolfach 1979. Proceedings 1979. Edited by D. Kölzow. XV, 573 pages. 1980.

Vol. 795: Séminaire d'Algèbre Paul Dubreil et Marie-Paule Malliavin. Proceedings 1979. Edited by M. P. Malliavin. V, 433 pages. 1980.

Vol. 796: C. Constantinescu, Duality in Measure Theory. IV, 197 pages. 1980.

Vol. 797: S. Mäki, The Determination of Units in Real Cyclic Sextic Fields. III, 198 pages. 1980.

Vol. 798: Analytic Functions, Kozubnik 1979. Proceedings. Edited by J. Ławrynowicz. X, 476 pages. 1980.

Vol. 799: Functional Differential Equations and Bifurcation. Proceedings 1979. Edited by A. F. Izé. XXII, 409 pages. 1980.

Vol. 800: M.-F. Vignéras, Arithmétique des Algèbres de Quaternions. VII, 169 pages. 1980.

Vol. 801: K. Floret, Weakly Compact Sets. VII, 123 pages. 1980.

Vol. 802: J. Bair, R. Fourneau, Etude Géometrique des Espaces Vectoriels II. VII, 283 pages. 1980.

Vol. 803: F.-Y. Maeda, Dirichlet Integrals on Harmonic Spaces. X, 180 pages. 1980.

Vol. 804: M. Matsuda, First Order Algebraic Differential Equations. VII, 111 pages. 1980.

Vol. 805: O. Kowalski, Generalized Symmetric Spaces. XII, 187 pages. 1980.

Vol. 806: Burnside Groups. Proceedings, 1977. Edited by J. L. Mennicke. V, 274 pages. 1980.

Vol. 807: Fonctions de Plusieurs Variables Complexes IV. Proceedings, 1979. Edited by F. Norguet. IX, 198 pages. 1980.

Vol. 808: G. Maury et J. Raynaud, Ordres Maximaux au Sens de K. Asano. VIII, 192 pages. 1980.

Vol. 809: I. Gumowski and Ch. Mira, Recurences and Discrete Dynamic Systems. VI, 272 pages. 1980.

Vol. 810: Geometrical Approaches to Differential Equations. Proceedings 1979. Edited by R. Martini. VII, 339 pages. 1980.

Vol. 811: D. Normann, Recursion on the Countable Functionals. VIII, 191 pages. 1980.

Vol. 812: Y. Namikawa, Toroidal Compactification of Siegel Spaces. VIII, 162 pages. 1980.

Vol. 813: A. Campillo, Algebroid Curves in Positive Characteristic. V, 168 pages. 1980.

Vol. 814: Séminaire de Théorie du Potentiel, Paris, No. 5. Proceedings. Edited by F. Hirsch et G. Mokobodzki. IV, 239 pages. 1980.

Vol. 815: P. J. Slodowy, Simple Singularities and Simple Algebraic Groups. XI, 175 pages. 1980.

Vol. 816: L. Stoica, Local Operators and Markov Processes. VIII, 104 pages. 1980.